I0812116

IMAGES
of Aviation

Michigan Aircraft Manufacturers

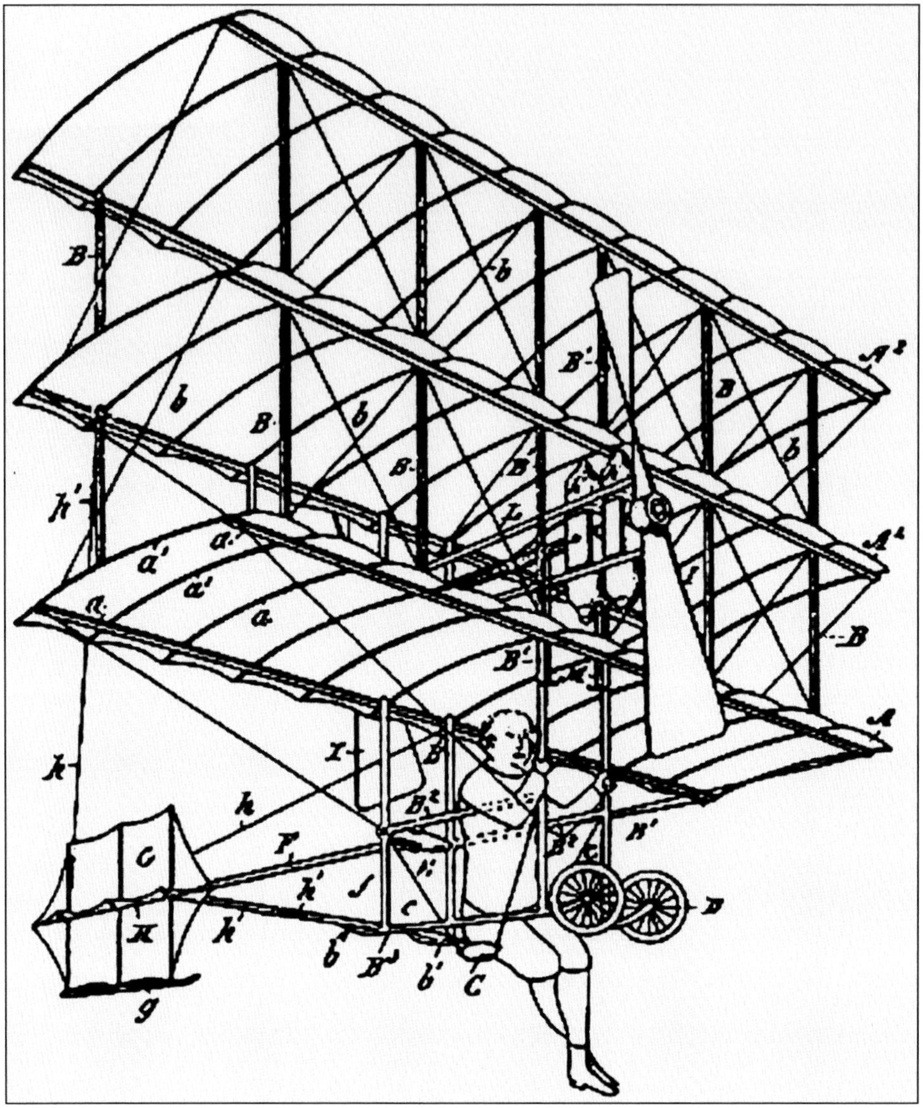

This is an illustration from a British patent application filed jointly by Octave Chanute, Chicago, and Augustus Herring, St. Joseph, Michigan, in 1896 for a heavier-than-air flying machine. A rather primitive design that used weight-shifting for control, it was based on the Chanute-Herring triplane glider flown by Herring in the summer of 1895. The patent, No. 15,221, was issued in June 1897. See page 11 for more information.

On the cover: Talbert "Ted" Abrams of Lansing, Michigan, is seen in front of the Abrams P-1 *Explorer*, a high-altitude photo-mapping airplane that he designed and built in 1937. The array of cameras displayed on the wing represents the wide variety that were used on photo-mapping missions. The large camera at the left is the Abrams model C-3 precision mapping camera, built by the Abrams Instrument Corporation. It used roll-film that could take 650 exposures in a nine-by-nine-inch image size. Abrams is holding a model of the proposed Abrams PC-4, a pressurized version of the *Explorer*, that was never built. See page 99 for more information. (Courtesy Abrams Aerial Survey Company.)

IMAGES
of Aviation

MICHIGAN AIRCRAFT MANUFACTURERS

Robert F. Pauley

ARCADIA
PUBLISHING

ISBN 978-0-7385-5218-7

Published by Arcadia Publishing
Charleston, South Carolina

Printed in the United States of America

Library of Congress Control Number: 2007941898

For all general information contact Arcadia Publishing at:
Telephone 843-853-2070
Fax 843-853-0044
E-mail sales@arcadiapublishing.com
For customer service and orders:
Toll-Free 1-888-313-2665

Visit us on the Internet at www.arcadiapublishing.com

This book is dedicated to those unsung engineers, designers, and dreamers who created America's aircraft industry and who, over a span of less than 80 years, advanced the art of aircraft design and manufacturing from flimsy wood and fabric flying machines to rockets that put men on the moon.

Contents

ACKNOWLEDGMENTS

This book could not have been completed without help from a number of aviation enthusiasts, photographers, and historians who have helped the author in his quest to record Michigan's aviation history. Thanks must go to William T. Larkins, California, founder of the American Aviation Historical Society and author of the definitive book on Michigan's most famous airplane, the Ford Tri-Motor. Equal thanks must be given to Peter M. Bowers, now deceased, a fountain of aviation knowledge who was most generous in sharing his collection of airplane photographs with the author. Plus thanks go to Dan Hagedorn for supplying the author with historical records on Michigan-built airplanes and to Dave Ostrowski, former editor of *Skyways* magazine.

Closer to home, the author is indebted to a number of Michiganders, all of whom have a deep interest in Michigan's aviation history. They include Prof. Peter Altman, deceased, who played a major role in many of the events depicted in this book; Dr. William Ballard, specializing in aviation history of the Saginaw area; John Bluth, author of the Arcadia book on Stinson Aircraft; Gerard Pahl and William Painter of the Kalamazoo Air Zoo Museum; and so many others that it would take several pages to thank them all. However, the one person who inspired the author to pursue his interest in Michigan's aviation history was Stephen Hudek, who is no longer with us. He was generous to a fault in sharing his airplane photographs with others. His collection of over 100,000 negatives included many images that would have been lost had it not been for his perseverance. He is sorely missed by all who knew him. Unless otherwise noted, all the images in this book came from his collection.

But this book would not have reached fruition had it not been for the encouragement and support provided by my wife, Margaret. She passed away while I was completing the final chapters of this book, and I dedicate this book to her memory. Thank you, my dear Margaret, for understanding my second love, aviation history.

INTRODUCTION

From the earliest days of recorded history, man has envied the flight of birds, and Egyptian, Greek, and Roman mythology all contain tales of man's attempt to fly. That goal was not reached until 1783, when two Frenchmen sailed over the city of Paris in a Montgolfier hot-air balloon. The flight was witnessed by Benjamin Franklin, then the American ambassador to France, who later wrote "this experiment is by no means a trifling one. It may be attended with important consequences that no one can foresee." In the years that followed, balloon flights became a popular attraction in Europe and were used to deliver mail during the siege of Paris in 1870–1871.

The first balloon flight in America took place on January 9, 1793, when a French balloonist, Jean-Pierre Blanchard, went aloft over Philadelphia, sailed across the Delaware River, and landed in New Jersey. The takeoff was witnessed by Pres. George Washington along with Thomas Jefferson, James Madison, and other leaders of the American Revolution. During the 1800s, balloon flights became a common sight at fairs and exhibitions across America, and during the Civil War, both sides used tethered balloons for observation and artillery spotting. But balloons had their limitations, since they were at the mercy of prevailing winds. Eventually early experimenters attached a gondola equipped with a motor-driven propeller to an elongated gas bag, creating an airship that gave the aeronaut limited directional control.

But the ultimate goal was a fully-controllable, man-carrying, flying machine that would overcome the limitations of both balloons and airships. During the 1800s, numerous inventors around the world were trying to solve the dream of flight. Among the more prominent were Sir George Cayley and Hiram Maxim in England, Clément Ader in France, Otto Lilienthal in Germany, Alexander Mozhaisky in Russia, and in the United States, Prof. John Montgomery, Gustave Whitehead, Octave Chanute, and Samuel Langley. All were working towards one common goal, manned flight. Without a doubt, Lilienthal, a German engineer, was the most successful of those early experimenters. He applied a scientific approach to his study of gliders and kept detailed records of his flights. Photographs and articles about his birdlike gliders were widely published in American newspapers. Lilienthal had made over 2,000 successful flights, which were controlled by weight-shifting. He was planning to add a motor with a propeller to one of his gliders but unfortunately lost his life in 1896 when his glider stalled and crashed.

In Dayton, Ohio, brothers Orville and Wilbur Wright had read about Lilienthal's flights and had become intrigued by the challenge of powered-flight. They wrote to the Smithsonian Institution, requesting literature on the subject of flying machines and were referred to Octave Chanute in Chicago. Chanute, a French engineer born in 1832, had come to America at an early age and eventually became one of this country's leading civil engineers. He, too, had become interested in the challenge of flight and had engaged in lengthy correspondence with Lilienthal and other aviation pioneers. In 1894, he published a booklet, *Progress in Flying Machines*, that soon became the bible for many of the early inventors. Too old to build or test fly his own creations, Chanute hired younger men like Augustus Herring and Paul Avery to build and test fly the gliders he had designed.

The Wrights corresponded with Chanute and other inventors and read whatever literature was available on the subject, but most of their ideas were original. Their approach to the challenge was methodical, starting with simple kites, then progressing one step at a time with man-carrying gliders to eventually reach the ultimate goal of powered flight. They built a wind tunnel to test airfoils and discovered that the data for lift and drag that Lilienthal had published contained errors, so they designed their own airfoils. Their first kite, a biplane-type that incorporated wing-warping for lateral control, was built in 1899 and tested at Kitty Hawk, North Carolina, where the steady winds provided ideal conditions for conducting tests.

Over a period of three years, from 1900 to 1902, the brothers built and flight tested a series of man-carrying gliders, making over 1,000 successful flights. The breakthrough in controlled flight came in 1902 when they interconnected the rudder and wing-warping controls so that their glider was fully controllable in all three axes.

After returning to Dayton, the brothers began planning the final step in their quest for powered-flight. Having solved the aerodynamic challenges of controllability, they now planned to add an engine and propeller to their basic glider design. After an unsuccessful search to find an existing engine suitable for their needs, the ever resourceful Wrights designed and built their own 12-horsepower, four-cylinder engine. They also built two remarkably efficient propellers that were chain-driven, an idea inspired by their bicycle-building background. Those components were assembled onto a new airframe, which was slightly larger than the 1902 glider. The Wrights referred to this new machine, the world's first practical airplane, as the *Flyer.* The genius of the Wright brothers became apparent on December 17, 1903, at Kill Devil Hills, near Kitty Hawk, North Carolina. On that day, Orville and Wilbur made four successful flights in the *Flyer.* The longest flight covered 852 feet and lasted one second shy of a full minute. The Wrights had taken off from level ground in a fully-controllable, man-carrying flying machine. On that cold morning in December, the airplane was born.

One

The Pioneers 1898–1916

The first flight of a lighter-than-air balloon in Michigan took place on August 16, 1858, when Ira Thurston and a passenger made a balloon ascension from the city of Adrian. They landed safely in Riga, 17 miles east of Adrian, but as his passenger was climbing out of the balloon's basket, Thurston became entangled in the gas bag and was carried aloft towards Canada. His body was found several years later along the shores of Lake Erie. And in 1880, Prof. E. D. Hogan went aloft near Jackson carrying a farmer's wagon suspended below his balloon to demonstrate how a balloon could be used to transport cargo. By 1900, balloons had become a common sight at fairs and public gatherings all across the state of Michigan.

The first flight of a heavier-than-air flying machine in Michigan took place from the Detroit Fairgrounds on July 14, 1910, when Arch Hoxsey made a flight in a Wright *Flyer*. The following year Wright Exhibition, pilot Howard Gill made several flights before large crowds in Saginaw. Two months later, in July 1911, a six-day aviation meet was held at the Detroit Fairgrounds with contests for speed, distance, and altitude, plus races between aeroplanes and automobiles. In the years that followed, airplanes became an increasingly common sight in Michigan as professional exhibition flyers as well as individuals in homebuilt airplanes gave demonstrations at local events.

In December 1908, Orville and Wilbur Wright visited Detroit to help organize the Aero Club of Michigan and to select a site for a flying field. The formation of the Aero Club created a great deal of interest in flying and in the manufacturing of aircraft in Michigan. A number of prominent Detroiters, mostly associated with the automobile industry, were members of the Aero Club and offered financial support to potential aircraft manufacturers. But by 1916, the pioneering era was coming to a close and America was on the brink of entering the war in Europe. New airplanes were being developed both in this country and overseas, and the American aircraft industry was about to expand rapidly to support the war effort.

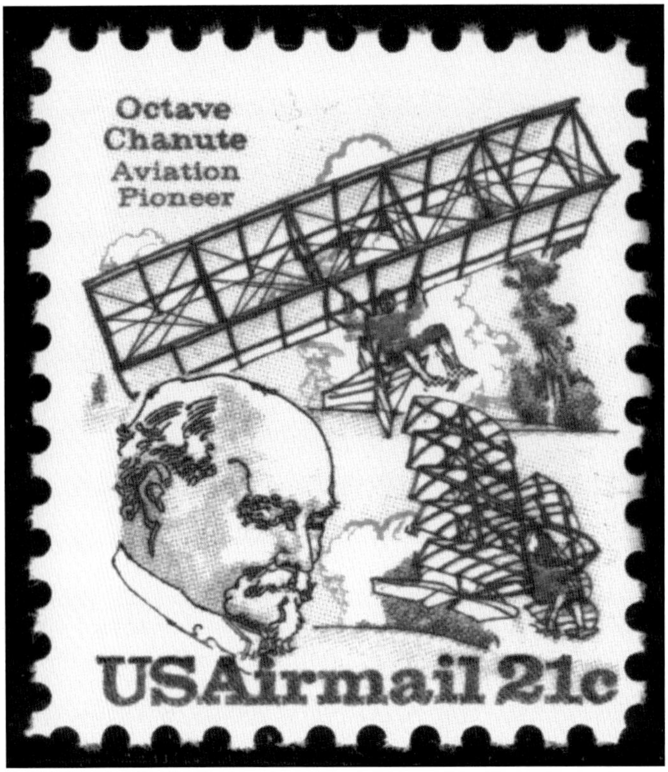

This 21-cent airmail stamp was issued by the U.S. Postal Department on March 29, 1979. It features a portrait of aviation pioneer Octave Chanute, and in the background, Augustus Herring of St. Joseph, Michigan, flying one of Chanute's gliders. Chanute and Herring collaborated on the design of several experimental gliders and tested them in 1895 on the Lake Michigan sand dunes near Gary, Indiana. See page 11 for more information.

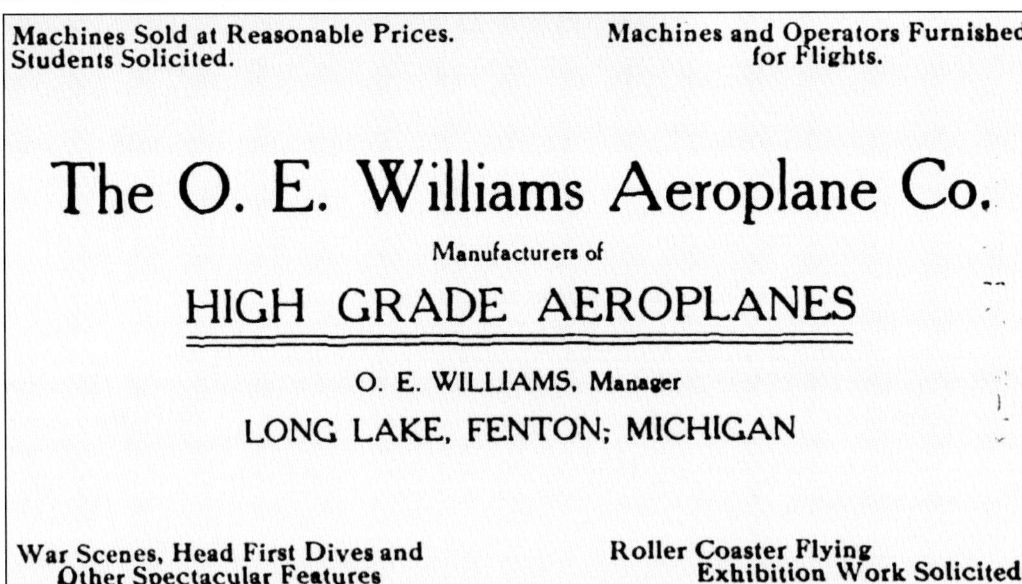

Machines Sold at Reasonable Prices.
Students Solicited.

Machines and Operators Furnished for Flights.

The O. E. Williams Aeroplane Co.

Manufacturers of

HIGH GRADE AEROPLANES

O. E. WILLIAMS, Manager

LONG LAKE, FENTON; MICHIGAN

War Scenes, Head First Dives and Other Spectacular Features

Roller Coaster Flying
Exhibition Work Solicited.

Osbert E. Williams's business card is seen here. In 1916, Williams ran a flying school in Fenton, Michigan, and manufactured the airplanes in which he taught his students to fly. The flying school had four instructors and his students were from the United States, Canada, Germany, Poland, Austria, and South America. Williams was known as "O. E." to his business associates but as "Bert" to his close friends. See page 15 for more information. (Courtesy Fenton Historical Society collection.)

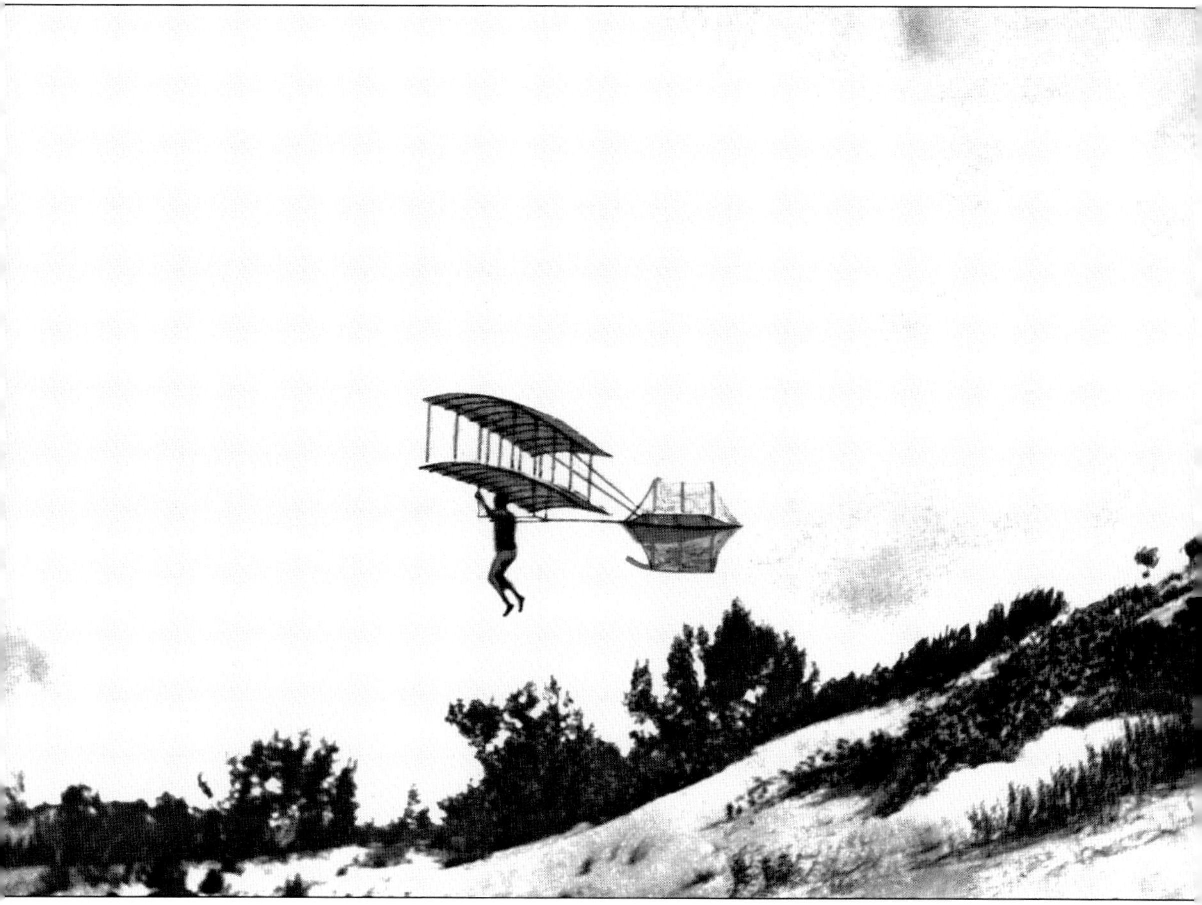

Augustus Herring was born in 1867 in Georgia and studied engineering at Stevens Institute. In 1894, he built a Lilienthal-type glider and made many successful flights with it. Herring was then hired by Samuel Langley, head of the Smithsonian Institution, Washington, D.C., to build a flying model of Langley's proposed *Aerodrome*, but Herring disagreed with Langley's concept and was fired. He moved to Chicago to work with Octave Chanute and built several man-carrying gliders designed by Chanute. In the summer of 1895, Herring and Paul Avery tested those gliders near Gary, Indiana. The Chanute-Herring biplane glider, with which Herring made over 700 flights, was the most successful (seen here). Herring later moved to St. Joseph, Michigan, and continued experimenting on his own. In October 1898, he mounted an air motor on a Chanute-type glider that drove dual propellers Herring claimed that he had made two successful "powered" flights with that machine but lacking reliable documentation, those claims have not been recognized. Herring died in Brooklyn, New York in 1927. (Courtesy Smithsonian National Air and Space Museum, Photograph Number 84-10696.)

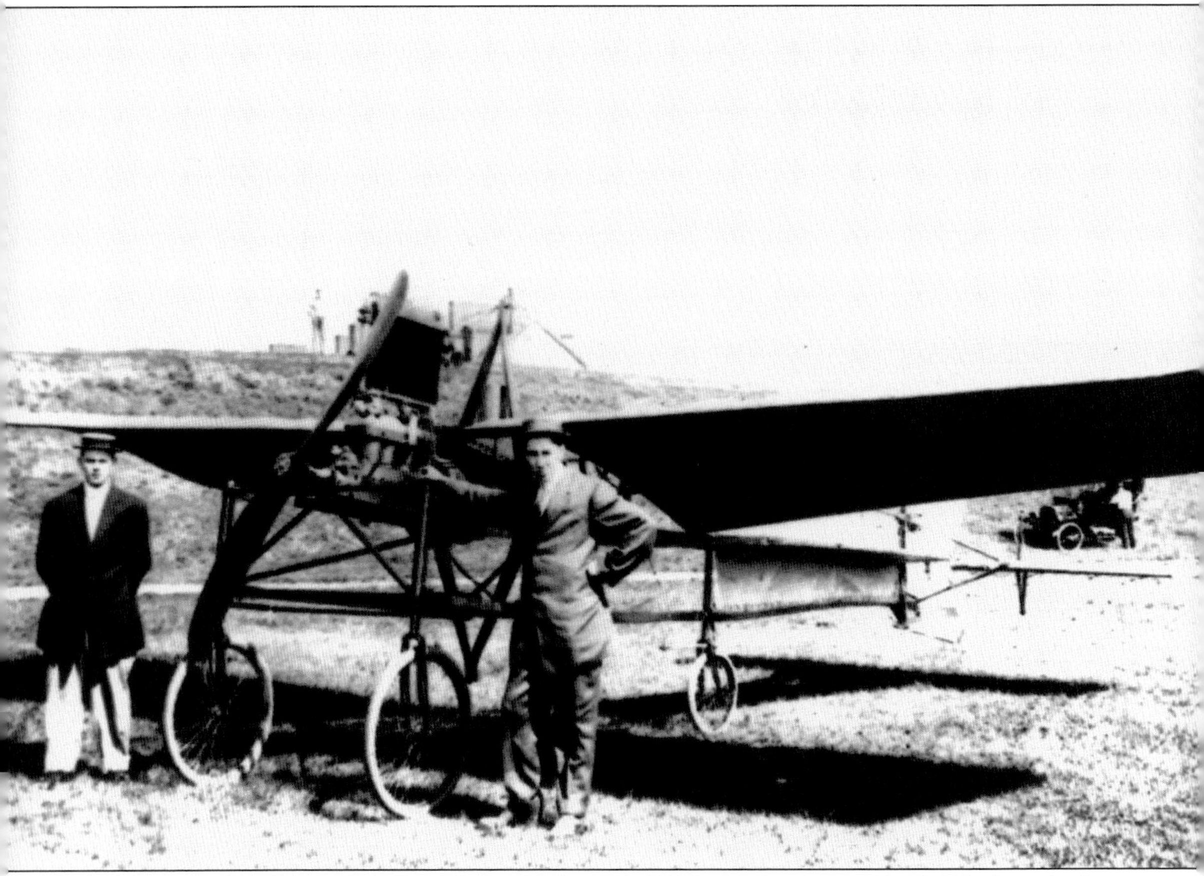

Edsel Ford was born in 1893, the son of automobile pioneer Henry Ford. Edsel worked part time in the Ford factory where, in 1909, he met a Ford employee, Charles Van Auken, who had purchased construction drawings for a French Blériot XI monoplane. Knowing that Edsel was deeply interested in aviation, his father asked several Ford employees to help Van Auken build parts for the airplane. Ford also donated a 28-horsepower Model T engine. The airplane was test flown by Van Auken on the Ford farm in Dearborn, Michigan. It rose a few feet above the ground, then settled back in a cloud of dust. Henry suggested the airplane needed more power and gave Van Auken a "hopped-up" engine. A second flight attempt was made from the Fort Wayne parade grounds in Detroit in 1910 (seen here). The airplane staggered into the air, but Van Auken lost control and crashed into a tree. He was slightly injured, but the airplane was destroyed. Some 15 years later, Edsel was to become deeply involved in aircraft manufacturing.

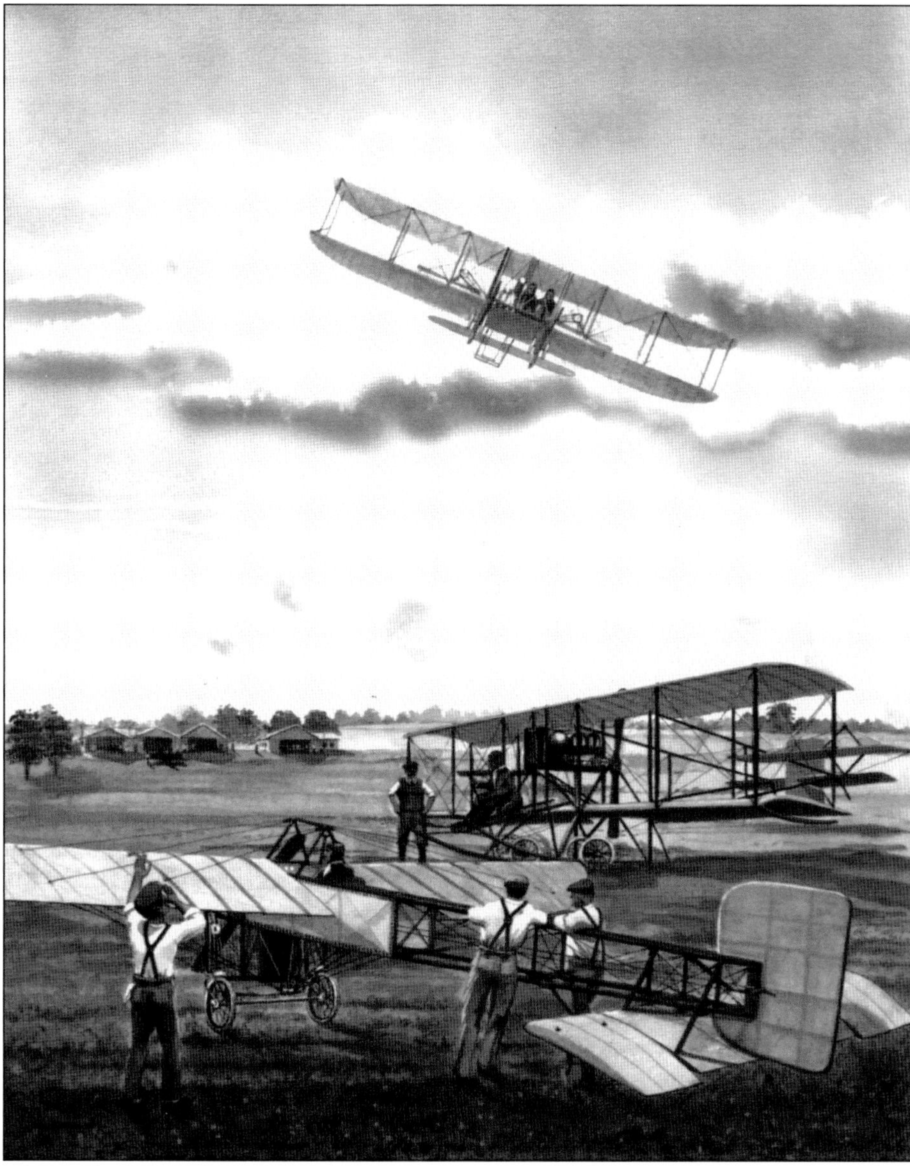

Clifford Brooks was president of the Brooks Manufacturing Company of Saginaw, Michigan. Founded in 1901, the company sold precut boat kits. Since early airplanes were of wooden construction, Brooks decided to manufacture airplane kits. Knowing little about airplanes, he hired engineer Edward Armstrong to design one for the Brooks Aeroplane Company. It closely resembled a French Blériot XI, except it was powered by a 60-horsepower Roberts engine. The single-seat prototype (seen here in the foreground) first flew in March 1911. Contemporary reports state that 22 kits were sold, but it is not known how many of those were actually built. Prices ranged from $25 for a basic kit (that included blueprints, instructions, and full-size templates) to $3,000 for a complete, ready-to-fly airplane. In 1912, Brooks hired E. R. Hamilton to design a Farman-type biplane, the model A (seen here in the background) that was sold to Hillary Beachy. Brooks got out of the airplane business in 1915 but continued selling boat and furniture kits until 1938. The airplane shown in flight is not a Brooks aeroplane. (Illustration by Lynn Morgan; courtesy Dr. William Ballard.)

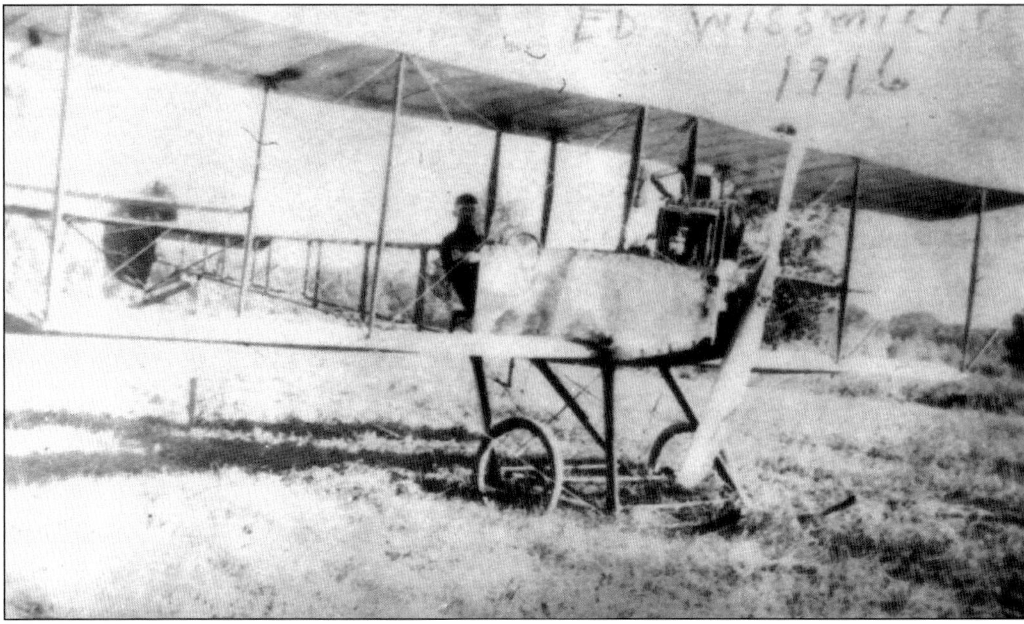

Don McGee was born in Ohio in 1896, and the family later moved to Saginaw, Michigan. As a teenager, McGee built and flew gliders but always wanted to build an airplane. That opportunity came in 1915, when the Brooks Aeroplane Company closed down. McGee bought leftover parts and assembled a tractor biplane with a Maximotor engine. Completed in June 1915, he taught himself to fly it and then began doing exhibition flying (seen above). McGee sold that airplane and in 1916 built a two-seat biplane with an OX-5 engine (seen below). Later that year, he was hired as a flight instructor by the O. E. Williams Flying School, Fenton, Michigan. In early 1917, he was hired as a civilian flight instructor at Selfridge Field, Mount Clemens, Michigan. On October 10, 1917, while on a solo flight, his airplane was seen in a steep dive into Lake St. Clair. McGee, only 21 years old at the time, died in the crash. (Courtesy archives of Michigan.)

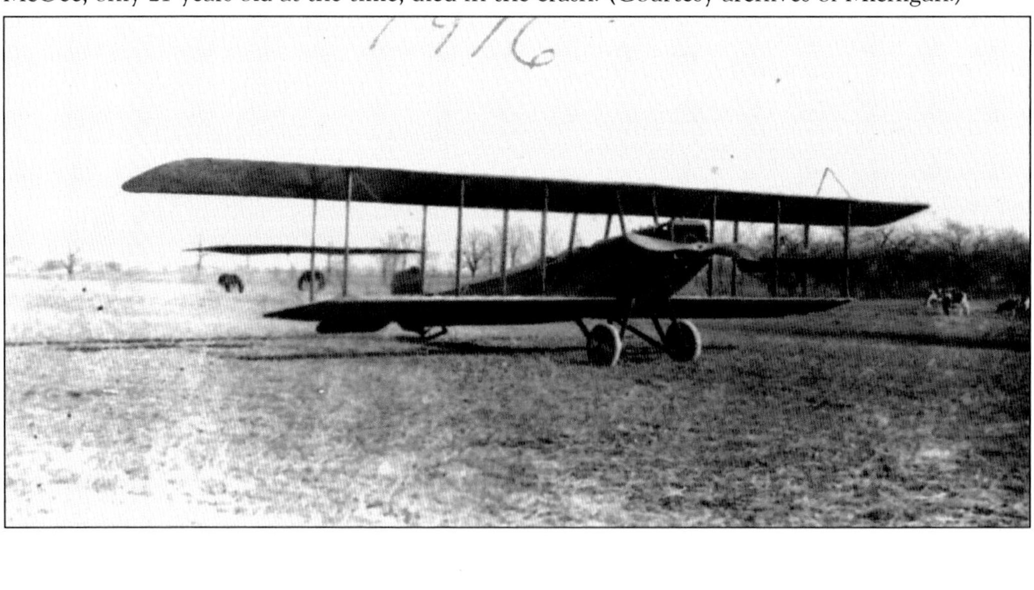

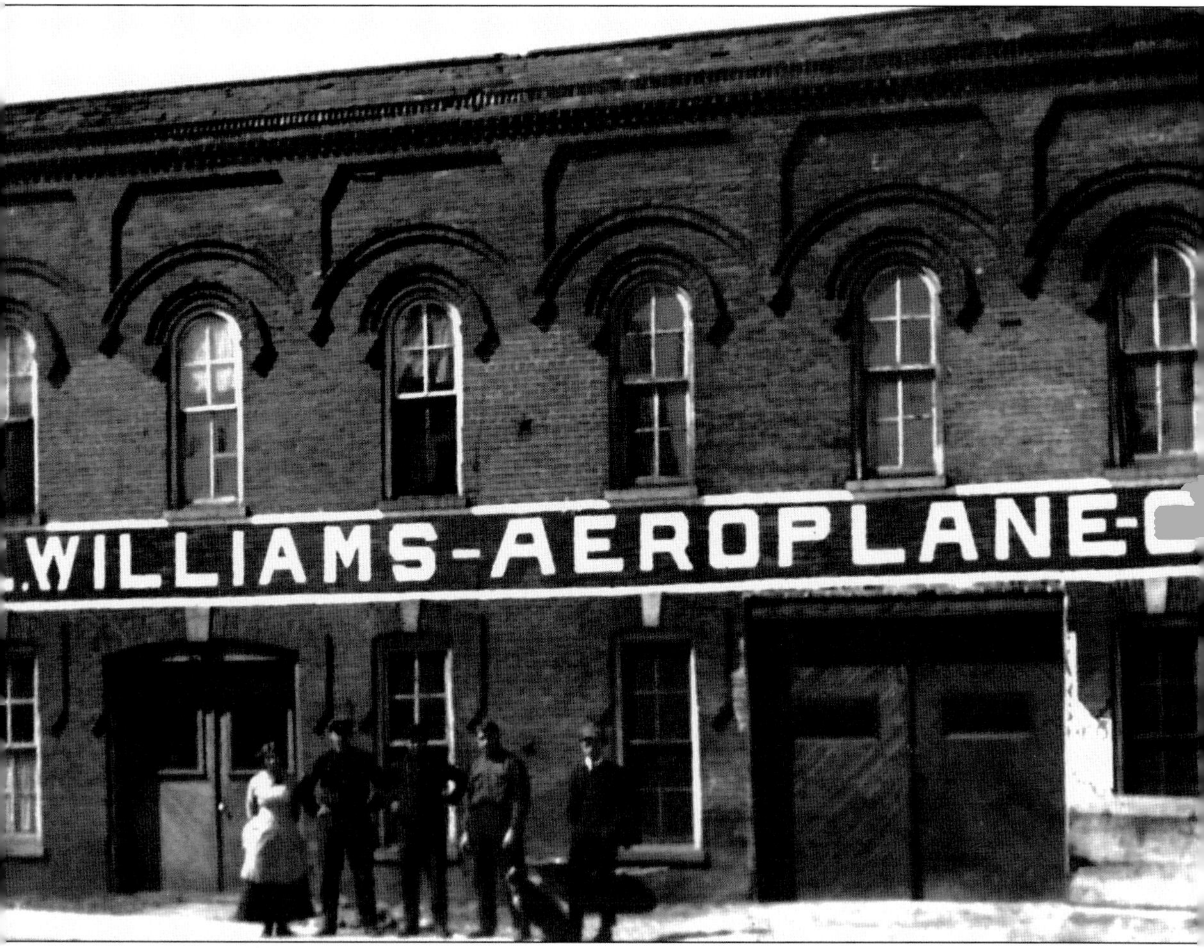

Osbert E. Williams was born near Webster, Michigan, in 1875 and learned to operate steam-powered farm machinery. In 1909, he was hired to manage a steam power plant in Pennsylvania, where he saw his first airplane and became interested in flying. In 1912, he built an airplane and taught himself to fly. He later built a two-place biplane and used it for exhibition flying. In late 1914, Williams flew an exhibition in Flint, Michigan, liked the area, and opened a flying school in nearby Fenton. The Williams' flying field was located on the north side of Long Lake. In addition to flight training, his students worked in the Williams Aeroplane Company's shop in Fenton (seen here), building or repairing two-place pusher biplanes that were used for flight training. Williams also built a two-place tractor biplane powered by a 125-horsepower, Williams-designed V-8 engine. He sold the company in 1917 to the Flint Aircraft Company, who built 12 of the Williams tractor biplanes. Williams then moved to Alabama to open a new flying school but was killed in an airplane accident in October 1917. (Courtesy Fenton Historical Society collection.)

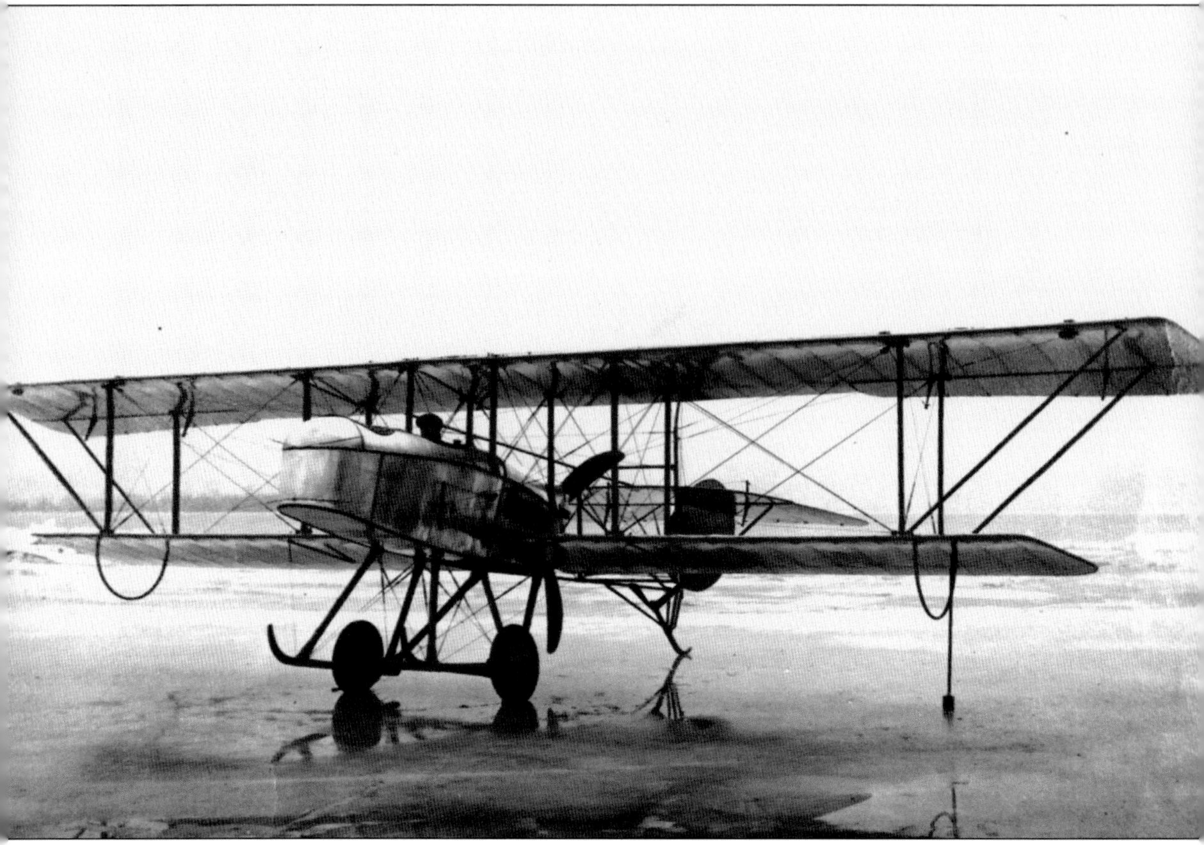

Alfred Verville was born in Atlantic, Michigan, in 1890 and by age 20 had completed a mail order course in engineering. In 1912, he was hired by the Curtiss Aeroplane Company in Hammondsport, New York, as a design engineer. He returned to Michigan in 1915 to form the General Aeroplane Company at 1507 East Jefferson Avenue, Detroit. His first design, the *Alpha*, was a two-place flying boat powered by a 90-horsepower OX-5 engine, that was sold to the U.S. Navy for pilot training. Verville then received a U.S. Navy contract to build a larger version, known as the *Beta*, but the contract was later canceled. During the winter of 1915–1916. Verville built a two-place fighter, the *Gamma* (seen here). It was taken to frozen-over Lake St. Clair for its maiden flight. Snow had been cleared to make a runway, but on takeoff, a wind gust blew the *Gamma* into a snow bank alongside the runway, and it crashed. Test pilot William Bonney was unhurt, but the *Gamma* was destroyed; Verville's General Aeroplane Company closed that summer. (Courtesy Fred Verville.)

Two

World War I
1917–1918

When the United States declared war on Germany on April 6, 1917, the U.S. Army Air Service was woefully unprepared for aerial combat in the skies over Europe. Although America was the first nation to purchase a military flying machine, the Wright model A of 1909, it had fallen far behind the advancements taking place overseas. The U.S. Army Air Service had only 55 airplanes in its inventory, mostly training aircraft, whereas the Europeans had developed large air fleets with long-range bombers and fast, well-armed fighters.

After war was declared, an Aircraft Production Board was formed to decide on a course of action. One major concern was the lack of a reliable aircraft power plant, so the board proposed that a standard aircraft engine be designed and put into production as quickly as possible. That task was assigned to engineers Jesse Vincent, from the Packard Motor Car Company, and Elbert Hall, from the Hall-Scott Car Company. In six days, they made preliminary drawings for an engine that could be built with either 4, 6, 8, or 12 cylinders. Packard built the prototype, a V-8, that ran for the first time on July 22, 1917. The board decided, however, to concentrate on the V-12 version, known as the Liberty, and over 20,000 were built by war's end.

Meanwhile the Bolling Mission was formed to select suitable European aircraft that could be mass-produced in this country. In June 1917, Maj. Raynal Bolling, along with several aeronautical engineers, visited European aircraft manufacturers and shipped back a number of sample aircraft to this country. Upon his return, Major Bolling recommended that America build a day bomber based on the British de Havilland DH-4 plus a large bomber, either the British Handley Page or an Italian Caproni.

The war in Europe had been going on for three years before America entered the conflict, so catching-up with the European aircraft designers had been a challenge. Despite many setbacks, the aircraft industry performed some remarkable feats of engineering and by the time the war ended had built over 14,000 airplanes, and aircraft manufacturing had grown from a handful of small, independent companies into a major industry.

PHOTOGRAPHS
OF
FISHER BODY CORPORATION
AEROPLANE DIVISION
DETROIT MICH
U.S.A.

SHOWING PART PLAYED IN HELPING WIN THE WAR:
JULY 19 1917 DECEMBER 24 1918

2005 AEROPLANES PRODUCED, AS FOLLOWS:

J-1 TRAINING PLANES -	400
DEHAVILAND-4 BATTLE PLANES -	1600
CAPRONI BI-PLANES -	5

200 SET DEHAVILAND-4 BATTLE PLANE SPARES
FIRST J-1 AEROPLANE SHIPPED JANUARY 15 1918
LAST J-1 AEROPLANE SHIPPED APRIL 17 1918
FIRST DEHAVILAND-4 BATTLE PLANE SHIPPED ON AUGUST 21 1918
LAST DEHAVILAND-4 BATTLE PLANE SHIPPED ON DECEMBER 24 1918
FIRST DEHAVILAND-4 BATTLE PLANE IN AIR ON JUNE 26 1918

MONTHLY PRODUCTION OF J-1 TRAINING AEROPLANES -		
	JANUARY	27
	FEBRUARY	109
	MARCH	172
	APRIL	92
		400

MONTHLY PRODUCTION OF DEHAVILAND-4 AEROPLANES -		
	AUGUST	100
	SEPTEMBER	204
	OCTOBER	501
	NOVEMBER	567
	DECEMBER	228
		1600

LARGEST PRODUCTION ANY ONE DAY - 40
LARGEST PRODUCTION ANY ONE WEEK - 150
LARGEST PRODUCTION ANY ONE MONTH - 567

DEHAVILAND-4 BATTLE PLANE NO. 24403 RIGHT SIDE SKELETON, LEFT SIDE FINISHED AND COVERED, TO BE PLACED IN THE SMITHSONIAN INSTITUTE, WASHINGTON, AS A PERMANENT EXHIBIT

TOTAL NUMBER OF EMPLOYEES - 4500

APPROXIMATE AMOUNT OF FLOOR SPACE DEVOTED TO AEROPLANE MANUFACTURE - 716,620 SQUARE FEET

FIRST UNIT OF AEROPLANE PLANT STARTED AUGUST 8 1917
COMPLETED IN 48 DAYS.

This title page is from a photograph album presented to Russell North, a test pilot for Fisher Body Corporation's Aeroplane Division. The page gives production figures showing Fisher Body's contribution to the war effort. With no prior aircraft manufacturing experience, Fisher Body built 2,005 complex airplanes over a 17-month period (note that the name of the British de Havilland DH-4 is misspelled). In 1919, Fisher Body bought the plant from the government. After General Motors Corporation acquired Fisher Body in 1926, the factory became the sole source of bodies for Cadillac automobiles. General Motors built 7.5 million automobile bodies in this plant before closing it in the 1980s. (Courtesy Kay North Weaver.)

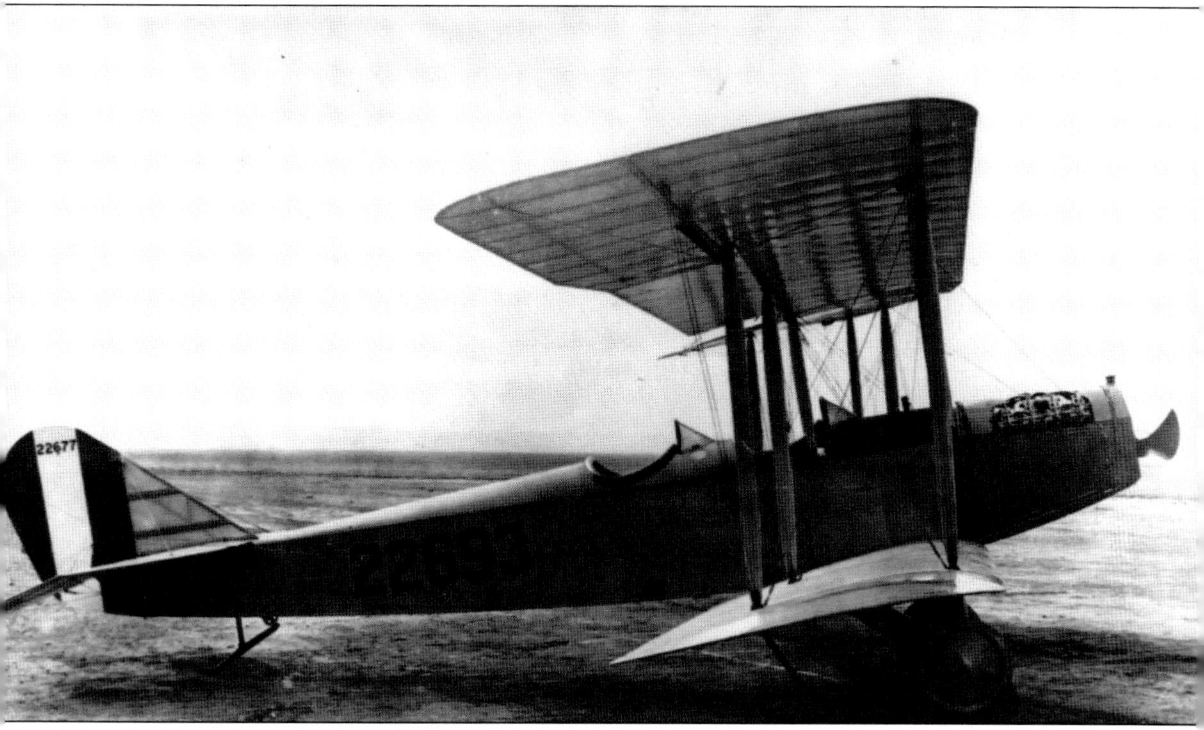

After the United States entered the war, there was a huge demand for aircraft in which to train new pilots. The standard U.S. Army Air Service trainer was the Curtiss JN-4 Jenny, but Curtiss was unable to keep up with the demand. Consequently the U.S Army Air Service selected a second primary trainer, the Standard J-1 (seen here), then in limited production in New Jersey. A contract was awarded to three manufacturers to build J-1s under license, and Fisher Body received a contract to build 400. The government built a new factory on Clark Street in Detroit and Fisher Body began building J-1s there even before the building was completed. Fisher Body's first J-1 was delivered to the U.S. Army Air Service in January 1918 and the 400th was delivered in mid-April of that year. The J-1 was a good airplane but was plagued with engine problems during its military career. In the postwar years, surplus J-1s were widely used by barnstorming pilots. (Courtesy Dave Ostrowski collection.)

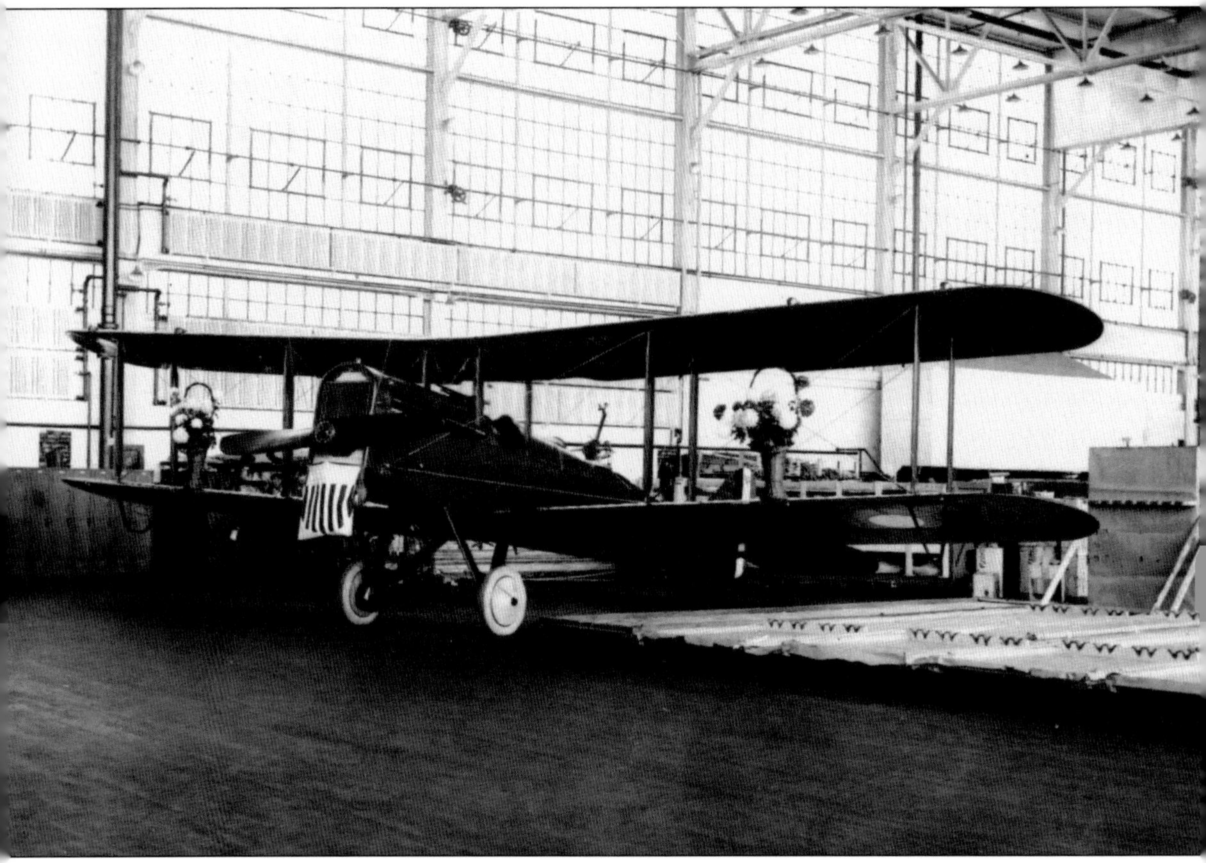

The Bolling Mission recommended that the British DH-4 be placed in production powered by the hew Liberty V-12 engine. A sample British airframe was delivered to the U.S. Army Air Service in August 1917 and tested in October, but there were numerous delays while the blueprints were being converted to American standards. Contracts had been awarded to three manufacturers, including Fisher Body in Detroit. After delivering the last Standard J-1 trainer Fisher Body began building DH-4s in the same factory. The first DH-4, now named the Liberty Plane, was completed in August 1918 and production increased rapidly, reaching a peak of 40 airplanes in one day. A total of 1,200 DH-4s were built by Fisher Body before production ended in December 1918. The first combat mission flown by Liberty Planes was on August 2, 1918, and over 1,000 had reached France by war's end. After the war, DH-4s were widely used by the U.S. Army Air Service and also by the U.S. Postal Service to deliver air mail. (Courtesy Kay North Weaver.)

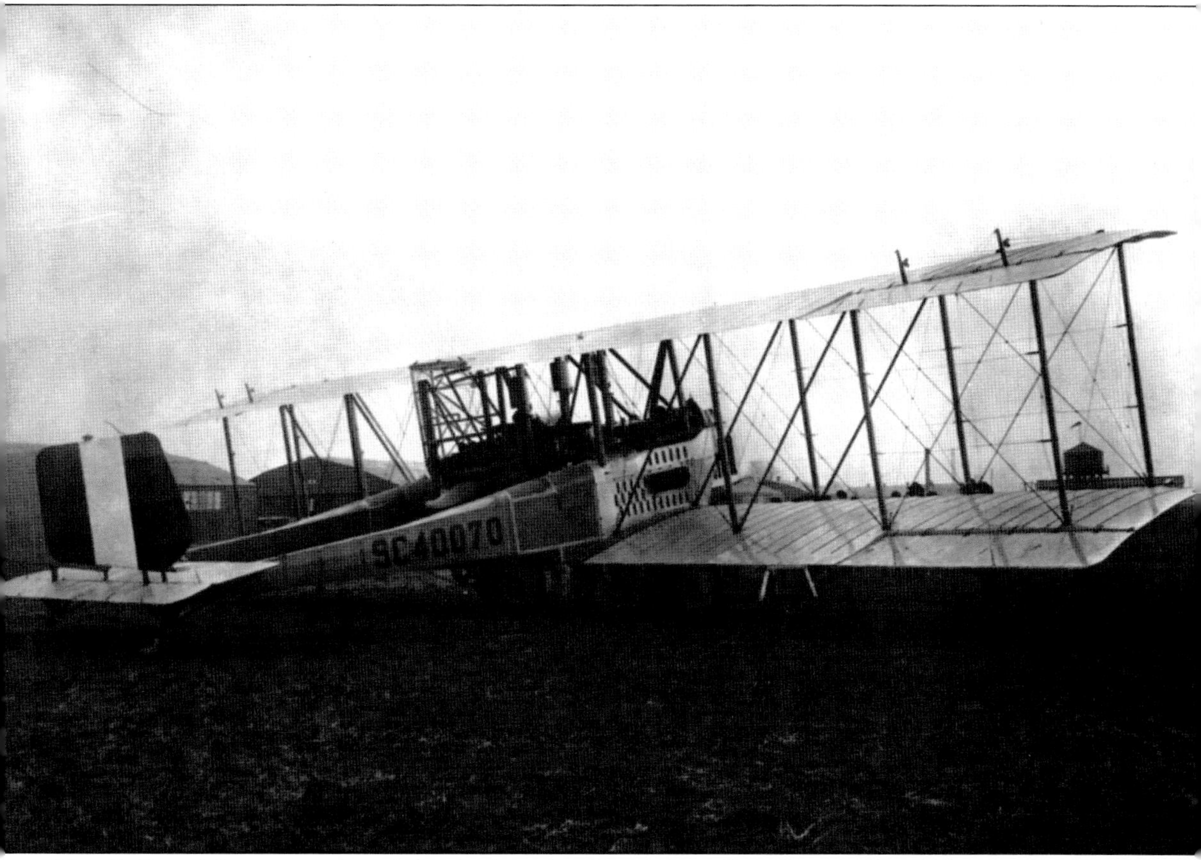

The Bolling Mission suggested that either the British Handley Page or the Italian Caproni bomber be built in this country. Two Capronis were shipped to America, a biplane and a huge triplane. After the triplane crashed, a contract for 1,000 twin-engine Caproni biplane bombers was awarded to the Standard Aircraft Company in New Jersey. But there were delays while the drawings were being converted from metric to inches and also adapting the airframe to the Liberty engines. Standard was also building a sample Handley Page bomber, further delaying progress on the Caproni bombers. As a result, in August 1918, Standard's Caproni contract was transferred to Fisher Body in Detroit, including two partially completed Caproni bombers. The first Caproni built by Fisher Body was shipped to McCook Field, Ohio, in November for static testing. Fisher built two more after the war, one in January 1919 and another in June. Fisher Body also completed Standard's two partially-built Capronis. All four were test flown at Morrow Field in Detroit and later delivered to the U.S. Army Air Service at McCook Field.

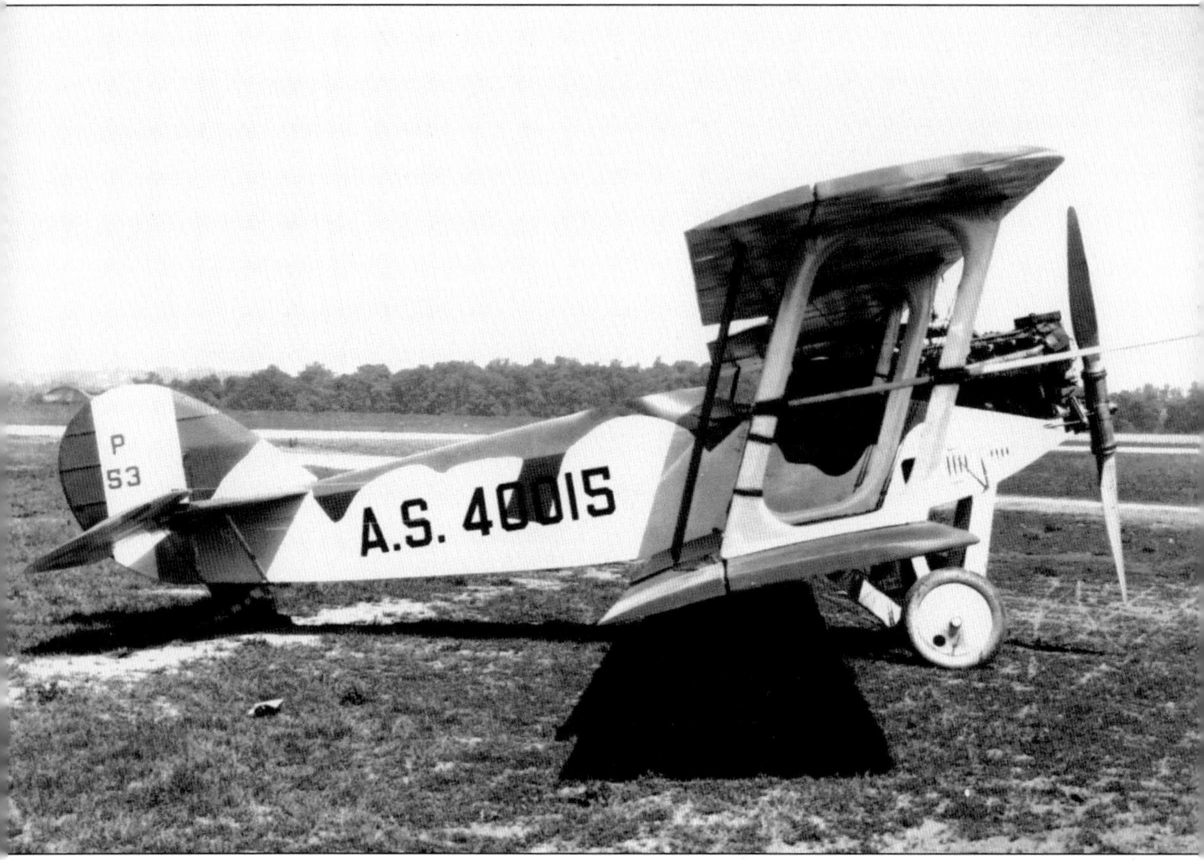

The U.S. Army Air Service's Engineering Division at McCook Field was responsible for developing new military aircraft but had not kept pace with progress overseas. As a result, an experienced French engineer, Capt. Georges LePere, was sent to America to design new airplanes. He was assigned to work at the Packard Motor Car Company in Detroit, where he designed this two-place fighter powered by a Liberty V-12 engine. The airplane had superior performance to the Liberty-powered DH-4s, just then entering production, so Packard Motor Car Company was awarded a contract to build 1,035. The first production LUSAC 11 (an acronym for LePere United States Army Combat) was completed in May 1918. Packard Motor Car Company had built 25, but when the war ended, the remainder of the contract was canceled. Two LUSACs had been shipped to France, but arrived too late to see combat. Three were built after the war plus one civil version for the postwar market. The U.S. Army Air Service used the LUSAC after the war to test superchargers, and one reached an altitude of 38,704 feet.

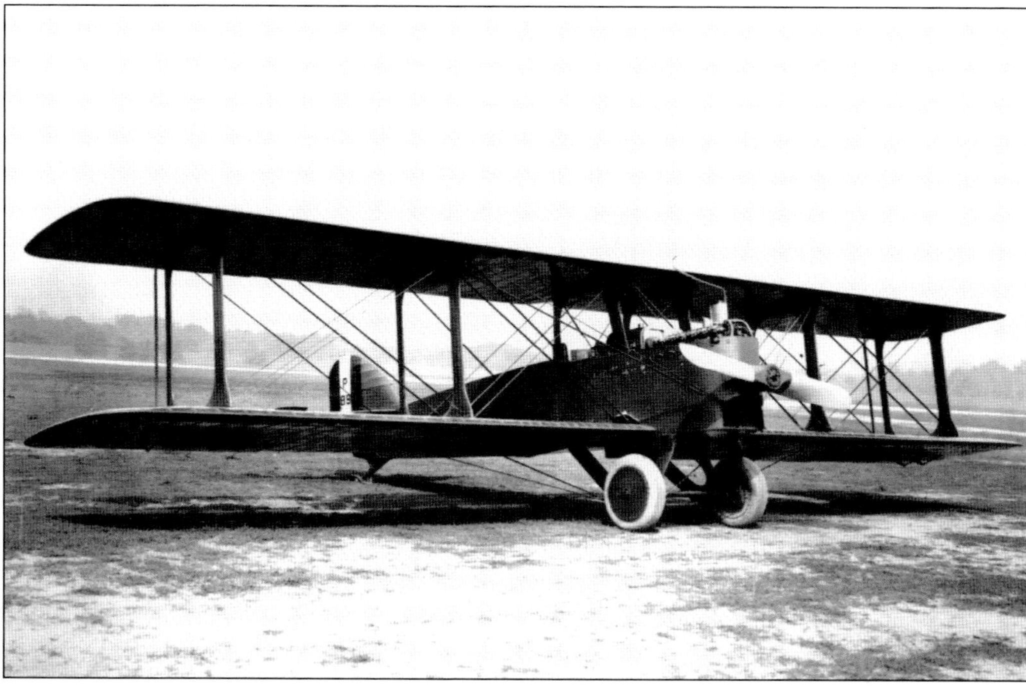

Captain LePere designed two other combat aircraft for the Air Service that were built by Packard after the war. The LUSAGH 11 (seen above) was a two-place side-by-side ground harassment airplane powered by a 16-cylinder King-Bugatti engine developing 450 horsepower. To protect the crew and engine from ground fire, the forward fuselage was lined with armor plate. LePere's other design was the LUSAO-11 (seen below), a twin-engine triplane designed for high altitude observation missions. It was powered by two 425-horsepower Liberty engines and carried a crew of three. Both airplanes were tested at McCook Field in 1919 and 1920.

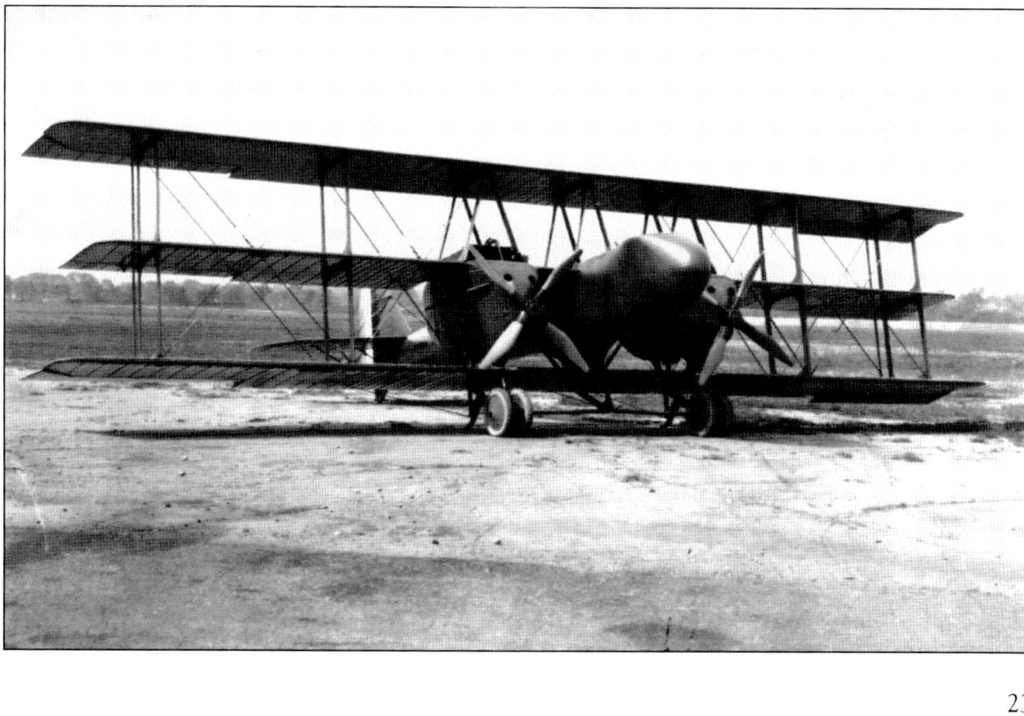

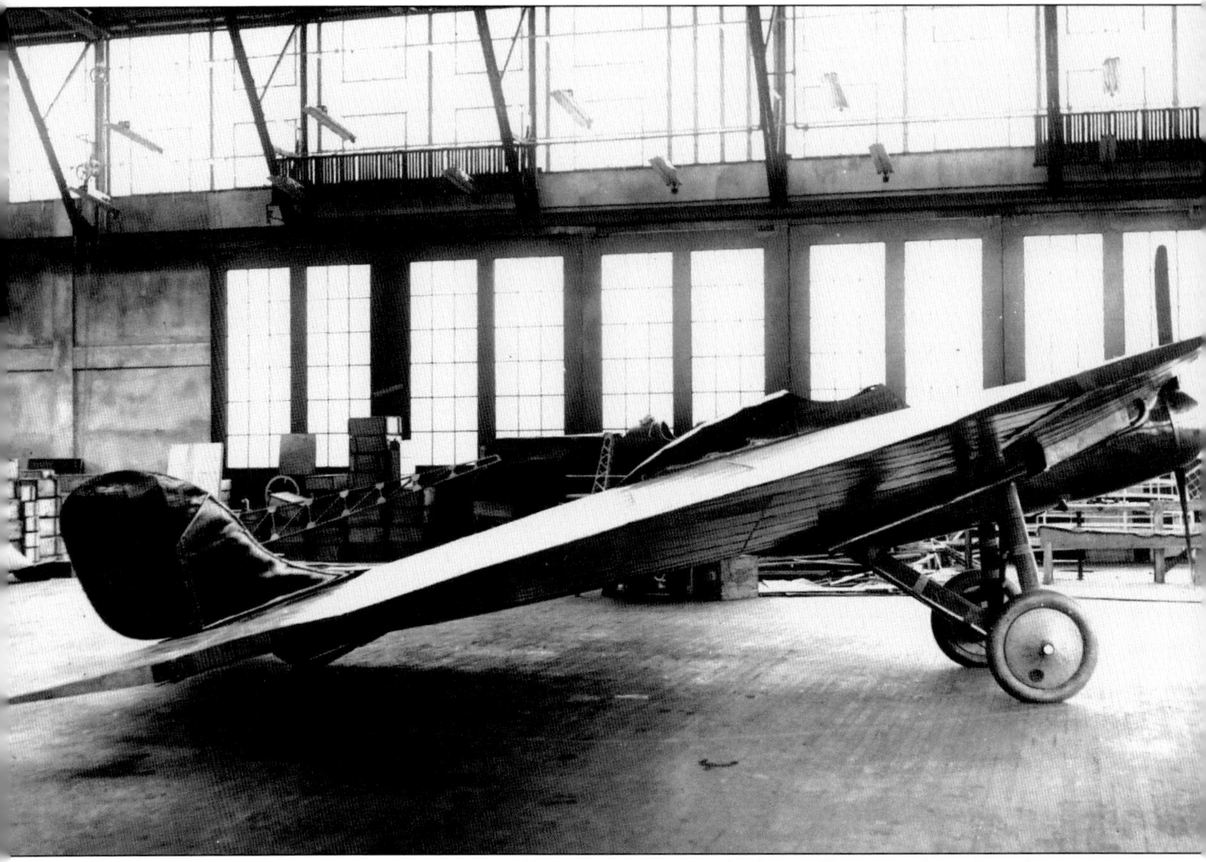

William Bushnell Stout was born in Illinois in 1880, and after college was hired as the aviation editor for the *Chicago Tribune*. He later founded *Aerial Age*, one of America's earliest aviation magazines. In 1914, Stout moved to Detroit to become chief engineer for the Scripps-Booth Automobile Company. In 1917, he was hired by Packard Motor Car Company's Aircraft Division to work on the Liberty engine program. Later that year, Stout was appointed technical advisor to the Aircraft Production Board in Washington, D.C. The board awarded Stout a contract to build a thick-winged monoplane featuring a blended wing-fuselage design. He returned to Detroit, obtained financial backing from Motor Products Corporation, and built this Stout *Batwing* in their shop. It was of wooden construction and was powered by a 150-horsepower Hispano-Suiza engine. The *Batwing* was delivered to McCook Field in Ohio in November 1918, and flown once by test pilot Lt. Jimmie Johnson. He commented that the airplane was too dangerous to fly because of the limited visibility from the cockpit, and the airplane was later scrapped.

Three

The Barnstorming Era 1919–1926

In the years following World War I, aircraft manufacturers had little incentive to build new airplanes when surplus military aircraft were available for a few hundred dollars. Many ex-military pilots bought an airplane, usually a Curtiss JN-4 Jenny, and embarked on a barnstorming tour, charging $1 or $2 for a 10 minute ride. Barnstormers made a valuable contribution to the public's growing interest in aviation by giving rural America a close-up view of an airplane and, for the daring few, a ride over their hometown.

Progress in aviation during the barnstorming period was dominated by military pilots whose record-breaking flights made headlines across the country. In May 1919, the U.S. Navy's NC-4 flying boat made the first aerial crossing of the Atlantic. In May 1923, two U.S. Army pilots flew nonstop from New York to San Diego in 26 hours and 50 minutes in a Fokker T-2. Also in 1923, a U.S. Navy pilot set a speed record of 266 miles per hour. In April 1924, four U.S. Army World Cruisers left Seattle on an around-the-world flight, but only two made it back to Seattle after a grueling 175 days. And in June 1924, a U.S. Army pilot crossed the continent in one day in a highly publicized "Dawn-to-Dusk" flight. That same year, a U.S. Army pilot set an altitude record of 38,704 feet in a Detroit-built Packard-LePere biplane.

But by the mid-1920s, the barnstormer's airplanes were rapidly wearing out, creating a growing need for replacements. In Michigan, manufacturers like Stinson, Buhl-Verville, and Hess introduced new airplanes that were a vast improvement over the old surplus military trainers. Furthermore some of the most significant advances in aircraft design were taking place in Michigan under the leadership of William B. Stout, an engineer, but more importantly, a man of vision. He designed the first all-metal airplane built in America and talked Henry Ford into manufacturing airplanes. The Stout-Ford Tri-Motors, built in Michigan, pioneered air travel in America and laid the groundwork for today's airline industry.

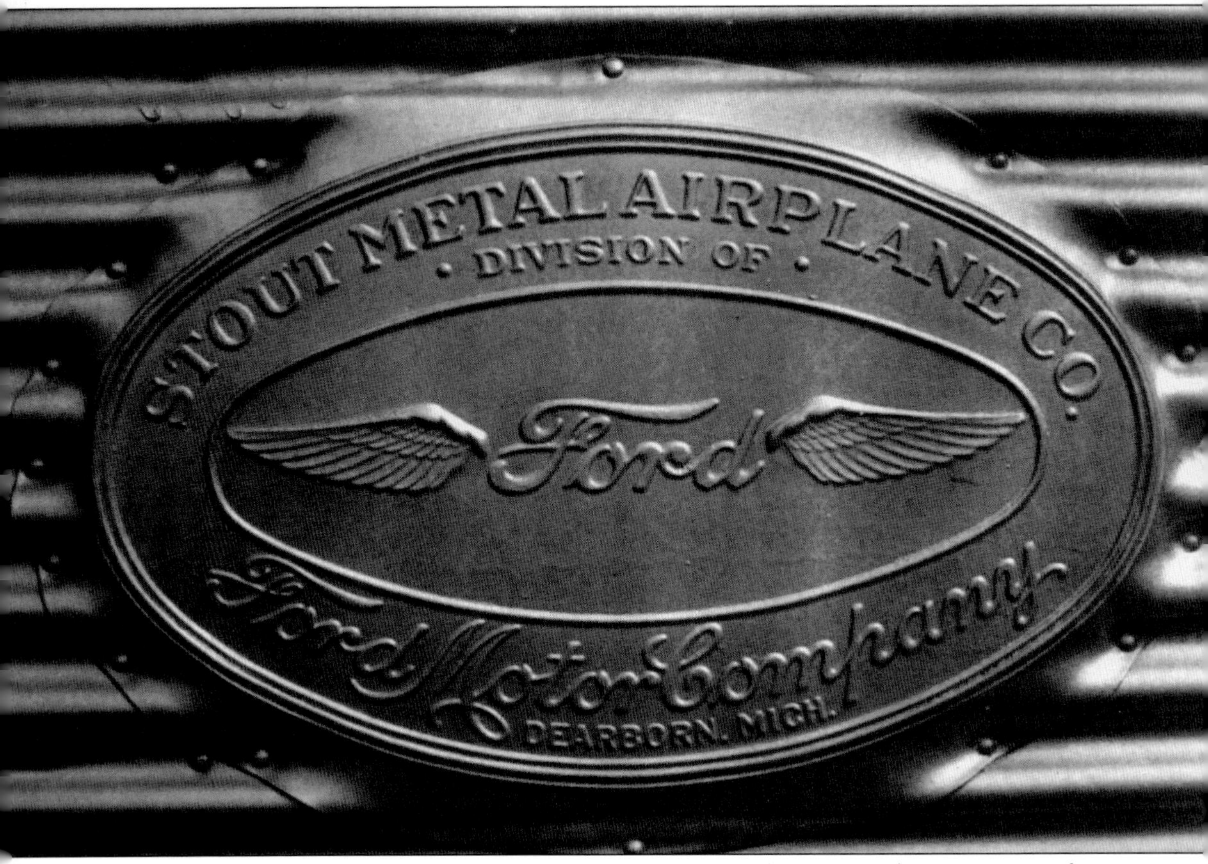

This aluminum medallion was located near the passenger door on Ford Tri-Motors, informing passengers boarding the aircraft that they were flying in an airplane manufactured by the Ford Motor Company in Dearborn. Approximately 18 inches across, the nameplate gave equal recognition to Bill Stout's contributions, stating that the airplane was built by the Stout Metal Airplane Company, a division of Ford.

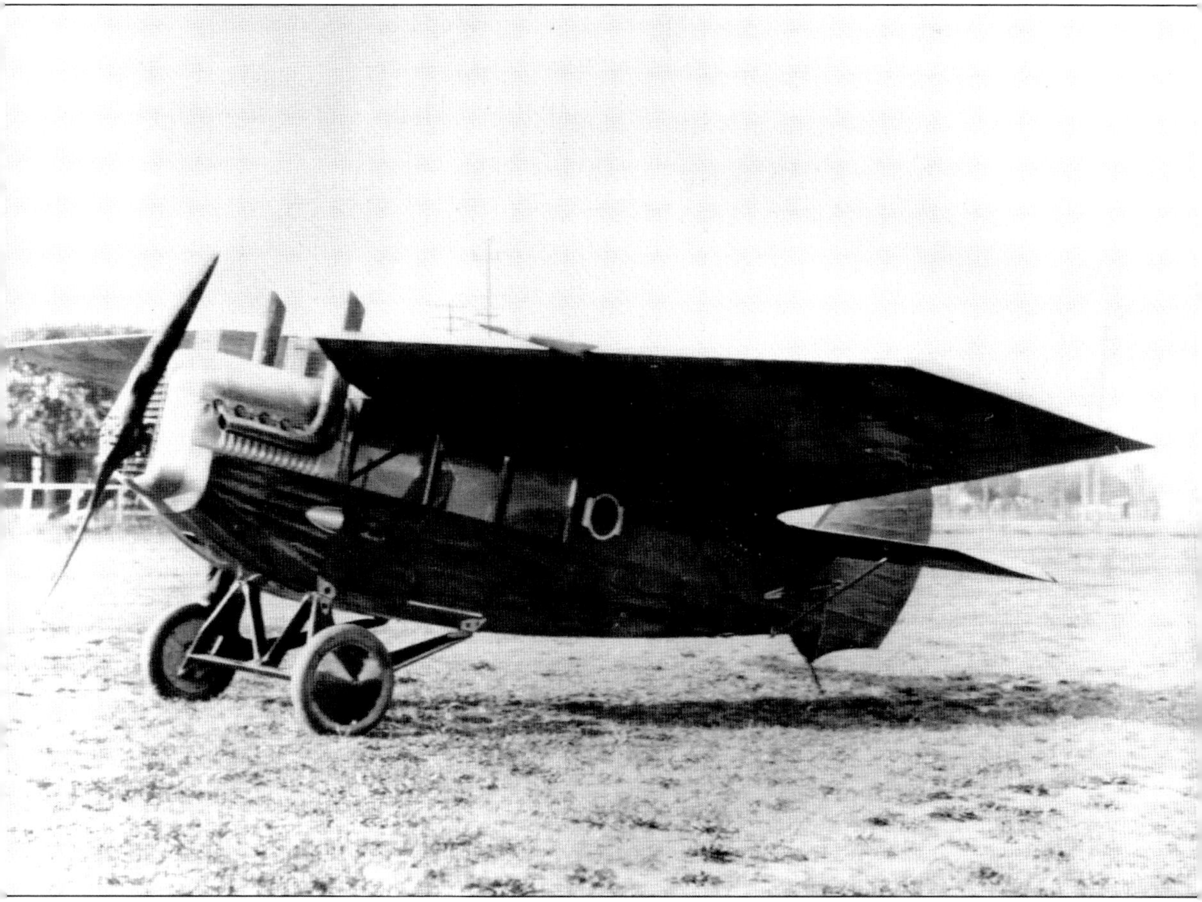

After building the *Batwing* for the U.S. Army Air Service in 1919, and later an airplane known as the *Cootie*, Bill Stout formed the Stout Engineering Laboratories to build commercial aircraft. He rented a building on Wellington Avenue in Detroit and with financial support from Robert Stranahan of the Champion Spark Plug Company, built this three-place, all-wood Batwing *Limousine*. It had an internally braced wing with a span of 36 feet and was powered by a 200-horsepower Packard engine. Stout made his own thin plywood to cover the airplane because plywood of the quality needed was not available at that time. The airplane was test flown by Bert Acosta at Packard Field in Detroit in late 1920. The Batwing *Limousine* generated considerable interest among military officials because of its ability to carry heavy loads.

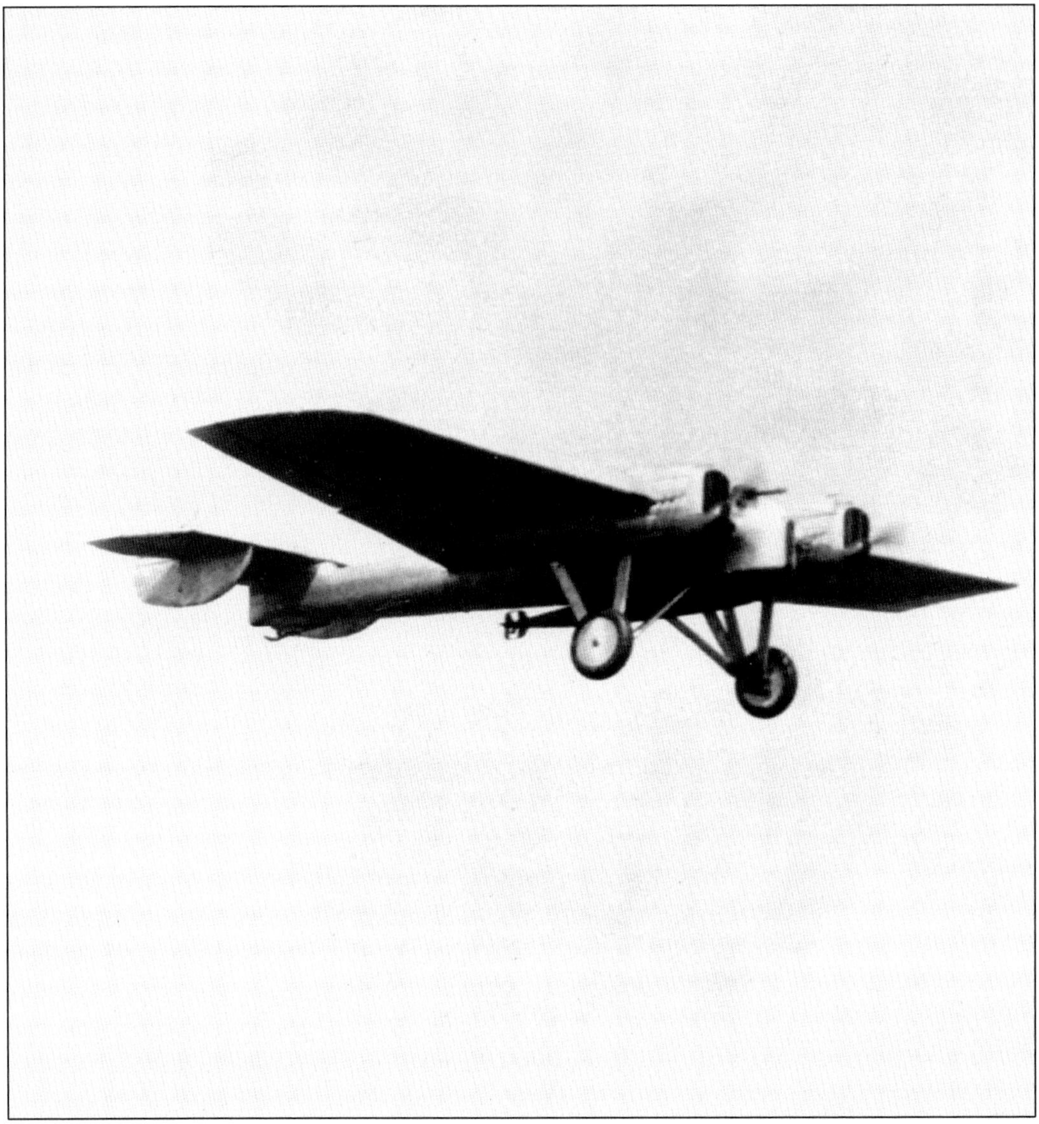

Based on Bill Stout's success in designing airplanes with thick, cantilever wings capable of carrying heavy loads, the U.S. Navy awarded Stout a contract to build three all-metal, twin-engine, Torpedo Bombers for $50,000 each. Stout hired George Prudden as chief engineer, and over a two-year period, they designed and built this Stout ST-1 in a shop on Beaubien Street in Detroit. The wing span was 60 feet, and it was powered by two 400-horsepower Packard engines. First flight was made by Eddie Stinson at Selfridge Field in Mount Clemens in April 1922. Later, at the official acceptance ceremony, the ST-1 was flown by a U.S. Marine Corps pilot who stalled the airplane while attempting to land. The pilot survived, but the ST-1 was a total loss, and the U.S. Navy canceled the remainder of Stout's contract.

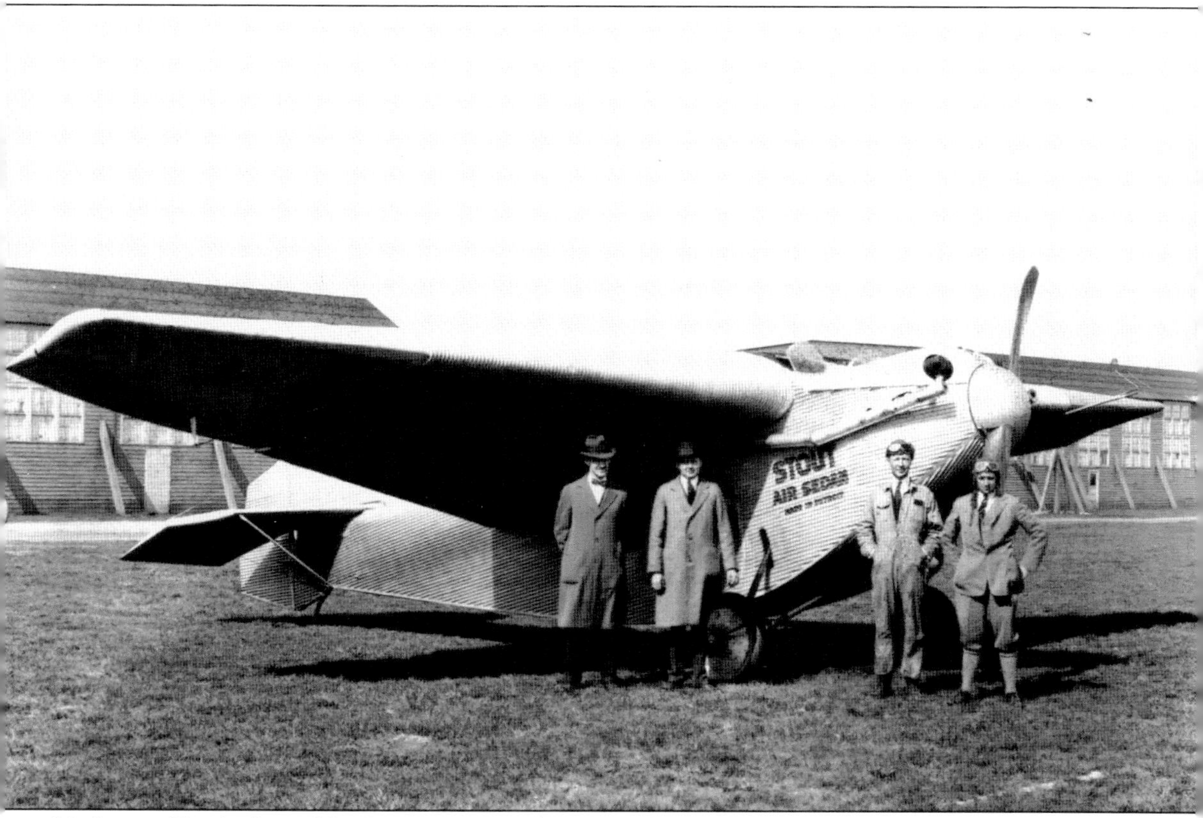

Undaunted by the loss of the ST-1, Stout was determined to continue building all-metal airplanes. In 1922, he personally typed over 100 letters addressed to prominent businessmen asking each to pledge $1,000 to promote aviation with only one promise, "That you will never see your money again!" Much to Stout's surprise he collected $128,000 from influential Detroiters, including Henry Ford and his son Edsel, and formed the Stout Metal Airplane Company on Beaubien Street in Detroit. There he built the Stout 1-AS Air Sedan, an all-metal, corrugated-skin, four-place monoplane powered by a 90-horsepower Curtiss OX-5 engine. First flight took place at Selfridge Field in Mount Clemens in February 1923 with Walter Lees at the controls (seen here, with Stout at far left). The airplane was underpowered, so to improve performance, Stout switched to a 150-horsepower Hispano-Suiza engine. Stout later offered rides in the Air Sedan to promote interest in Michigan's growing aircraft industry.

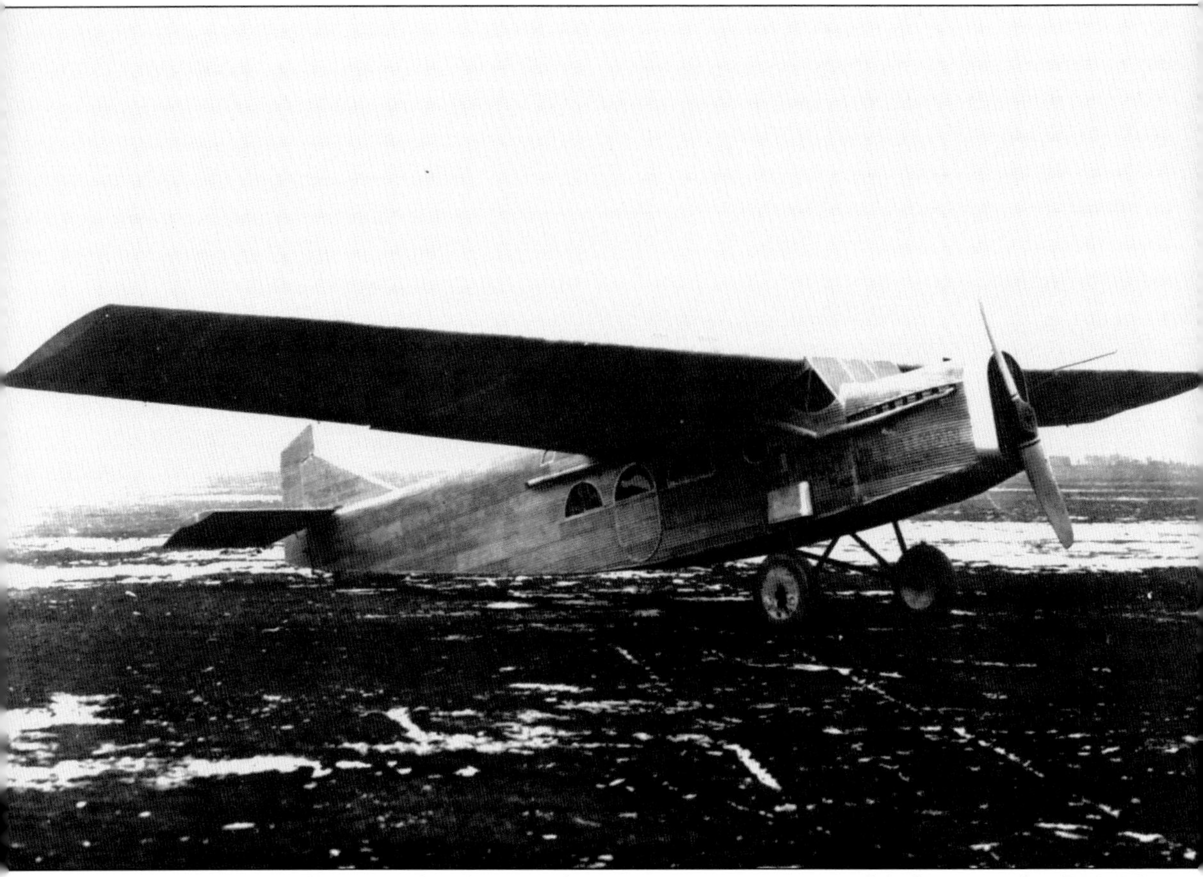

The U.S. Post Office introduced air mail service in 1919 using modified war-surplus DH-4M Liberty Planes. Trying to maintain a schedule with obsolete, open-cockpit biplanes was difficult and dangerous. Bill Stout believed that he could design a more efficient airplane using the same 400-horsepower Liberty engine that powered the DH-4M mail planes. Thus was born the eight-place, all-metal, Stout 2-AT Air Pullman, built in Stout's shop in Detroit. The airplane was an enlarged version of the *Air Sedan* with corrugated skin and cantilever wings. The prototype, named *Maiden Detroit*, was first flown by Walter Lees at Selfridge Field in Mount Clemens in April 1924. As an example of the casual approach to test flying in those early days of aviation, after a second test flight, Stout, his wife, and two friends, climbed aboard the 2-AT and went for a sightseeing flight over the city of Detroit.

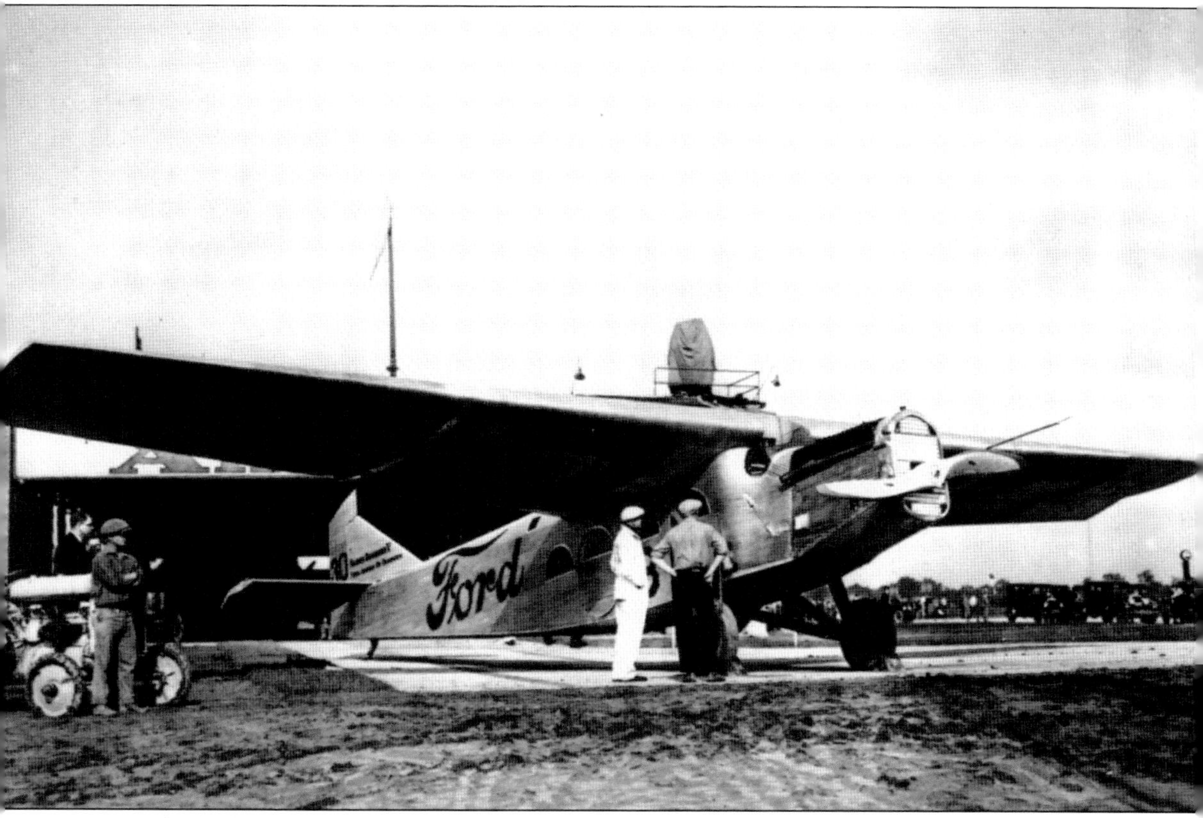

Henry and Edsel Ford closely followed the building and first flight of Stout's 2-AT Air Pullman, and they both saw a future in building airplanes as well as automobiles. Stout wrote a letter to Henry in which he mentioned that there were no major airports in the Detroit area, so Henry built a 600-acre airfield on his River Rouge property in Dearborn, including a hangar and a small factory building. In early 1925, Henry invited Stout to move his company from Detroit to that factory where Stout continued building 2-ATs, now named Air Transports. In April 1925, Henry bought a Stout 2-AT, and to demonstrate the practical use of air transportation, formed the Ford Air Transport Service. Ford's air service made daily flights between Dearborn and a Ford factory near Chicago, carrying freight and company mail. Ford later bought more 2-ATs and expanded the service to Cleveland and Buffalo. After passage of the Kelly Air Mail Act, the U.S. Post Office awarded Ford several government subsidized Contract Air Mail routes. The Ford Air Transport Service closed down in 1932 after having flown nearly two million miles.

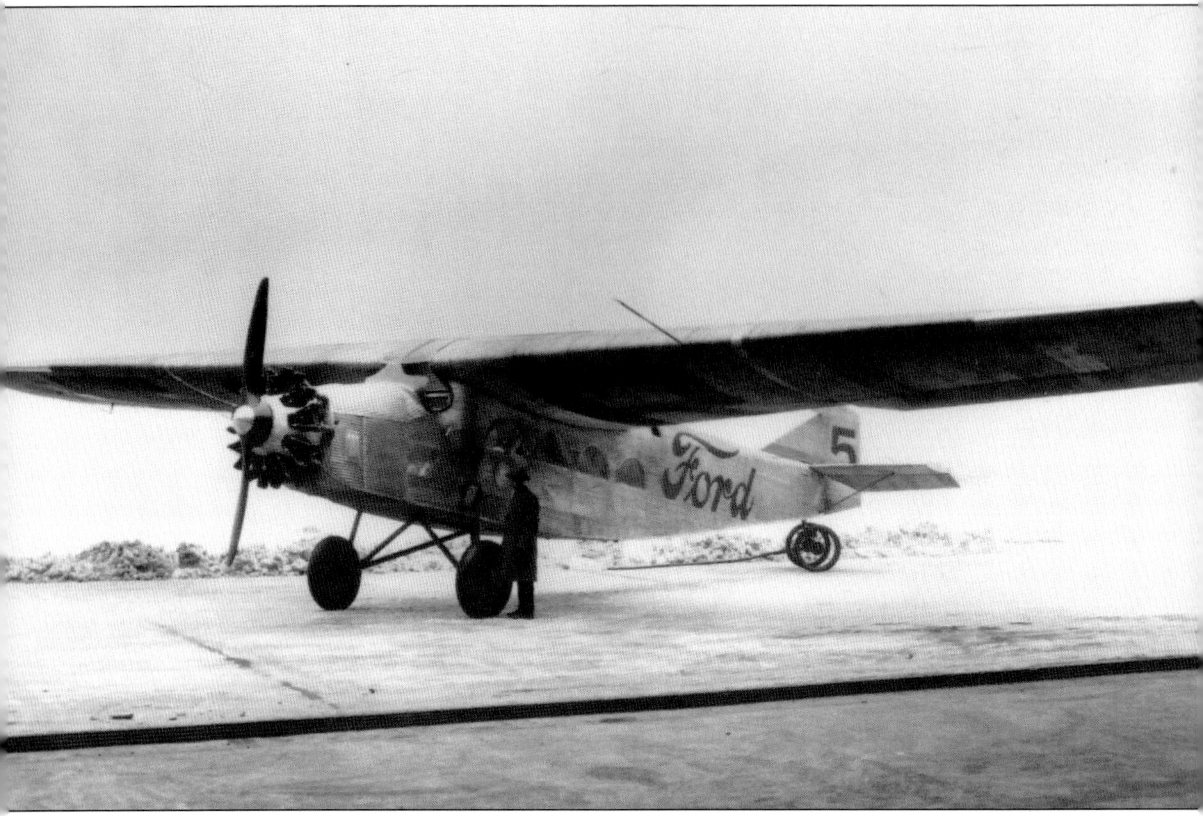

Eleven Ford-Stout 2-AT Air Transports were built, all powered by liquid-cooled, 400-horsepower Liberty engines except for one experimental model (seen here). It was tested with a 450-horsepower Pratt and Whitney Wasp air-cooled radial engine, but did not go into production. The Ford Air Transport Service operated six 2-ATs, and four were sold to Florida Airways. Florida Airways was formed by wartime aces Eddie Rickenbacker and Reed Chambers and operated a passenger and air mail service between Atlanta and Miami. One other 2-AT was sold to the John Wanamaker Department Store and displayed in their New York City store in October 1925. A 2-AT sold for $22,500. Despite their initial success, all 2-ATs were grounded by the U.S. Department of Commerce in August 1928 due to structural weakness in the wings, and all were scrapped.

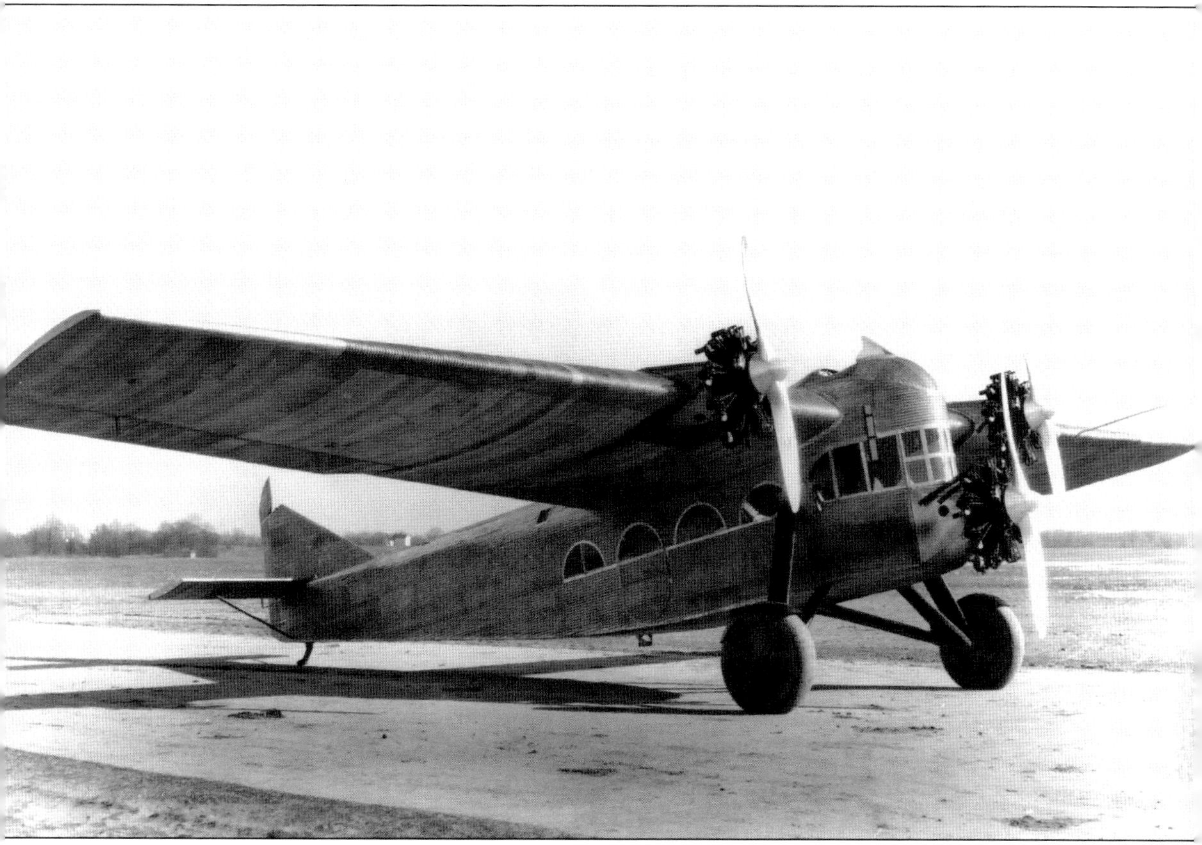

After Bill Stout moved his company to Ford Airport, he began building 2-AT Air Transports on an automobile-type production line. Henry Ford saw a future in manufacturing and selling airplanes, so on July 31, 1925, he bought out Stout's company and made it a division of the Ford Motor Company. Ford wanted to build all-metal transports but preferred multiengine types for safety. Stout told Ford that before moving to Dearborn he had been modifying a 2-AT into a tri-motor by replacing the single Liberty engine with three Wright J-4 radial engines. Stout's approach allowed him to build an experimental tri-motor at minimum cost, but the end result was an ungainly-looking airplane. Ford's chief test pilot Rudolph Schroeder flew the Ford-Stout 3-AT on November 25, 1925, but found the airplane so difficult to control that he refused to make a second flight. It was later flown by Larry Fritz who confirmed Schroeder's opinions. Henry Ford was not happy with Stout's makeshift tri-motor so the 3-AT was hidden away in Ford's hangar.

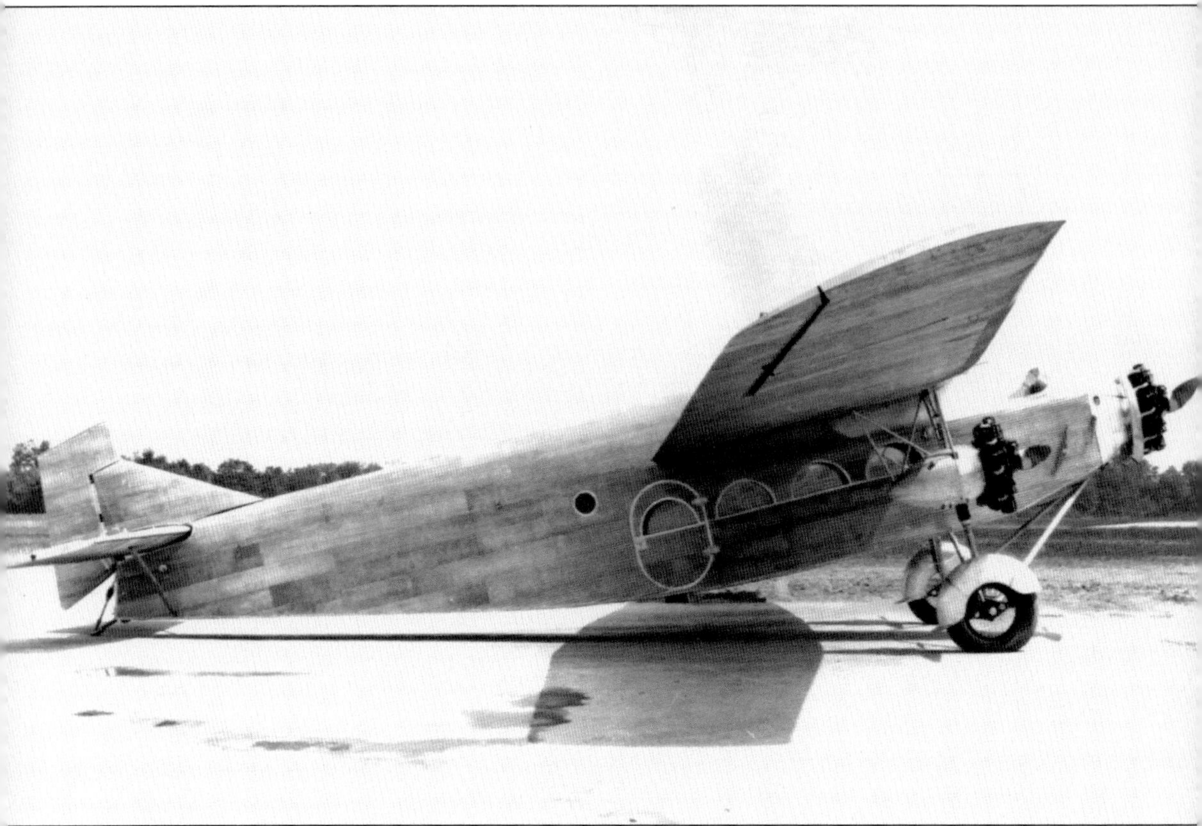

On January 16, 1925, the Ford-Stout factory and the Ford hangar burned to the ground, and the 3-AT tri-motor, plus several Air Transports, were destroyed. After the fire, Ford sent Stout on a nationwide lecture tour to promote Ford's aviation activities. Unknown to Stout, an aircraft design group had been formed in Ford's Laboratory Building under the leadership of Harold Hicks. The group, including aeronautical engineers Tom Towle and Otto Koppen, began designing a new tri-motor only a few days after the fire. Additional Ford employees were brought in later to assist with drafting work and building the airplane. The first true Ford Tri-Motor, the 4-AT-1, was test flown by Rudolph Schroeder on June 11, 1926. This was a remarkable engineering accomplishment. The group had designed, built, and test flown the prototype airplane in less than five months. Bill Stout was away on his lecture tour and had no input into the design. While the 4-AT-1 was being built a new and larger factory building was also built and plans were being made to put the new Ford-designed Tri-Motor into production.

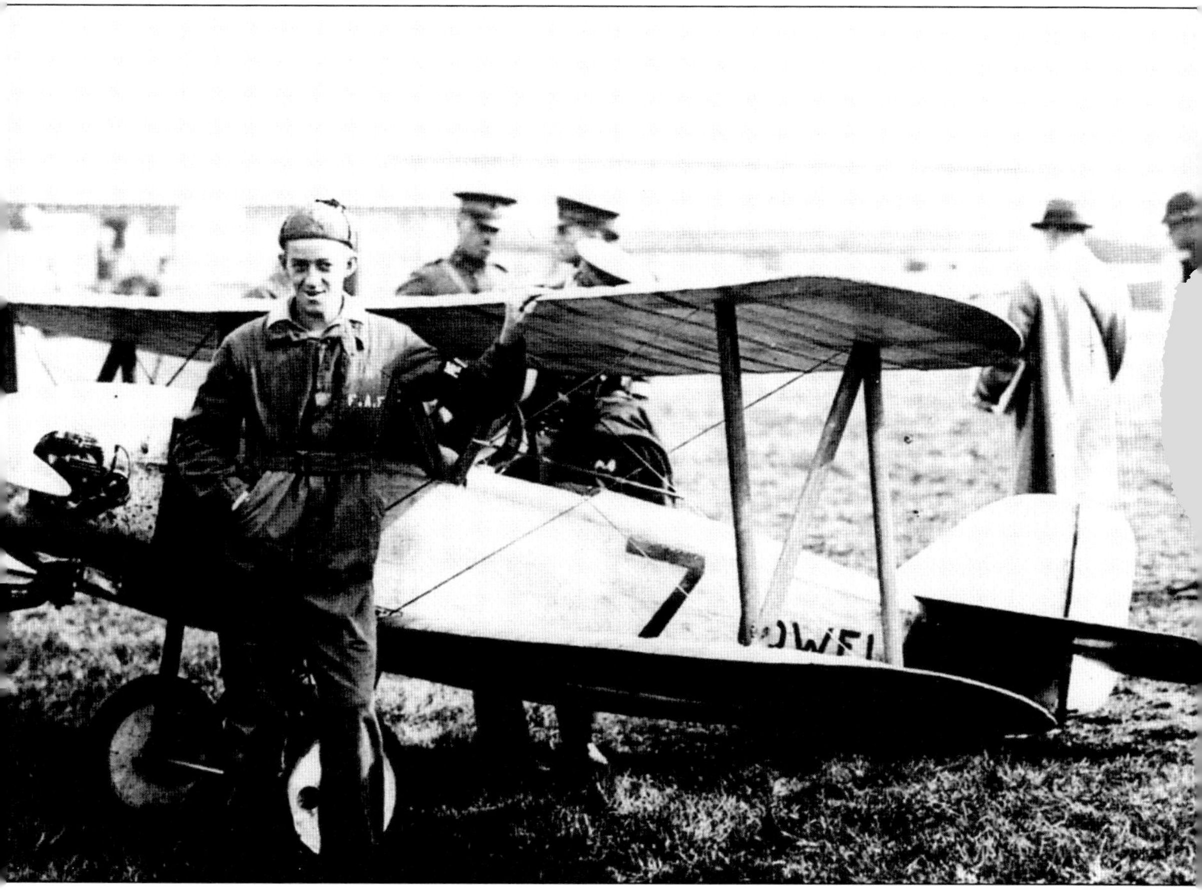

In 1920, the University of Detroit formed an Aeronautical Engineering Department under the direction of Prof. Clarence Powell. An Englishman, Powell had worked for Sopwith Aviation during World War I, and after the war came to America to teach aeronautical engineering. Powell was an advocate of a hands-on approach to teaching, so in 1924, he had his University of Detroit students design as well as build a diminutive biplane known as the *Powell Racer*. It had a wingspan of under 16 feet, weighed 310 pounds and was powered by a 32-horsepower Bristol Cherub engine. In October 1925, the airplane was entered in the Light Plane Races at Mitchell Field, Long Island. Piloted by Jerry Dack (seen here), the racer won the *Scientific American*, the *Aero Digest*, and the *Dayton Daily News* trophies. After its return to Detroit, the airplane was used for static testing by University of Detroit students and eventually destroyed. Professor Powell returned to England in 1926.

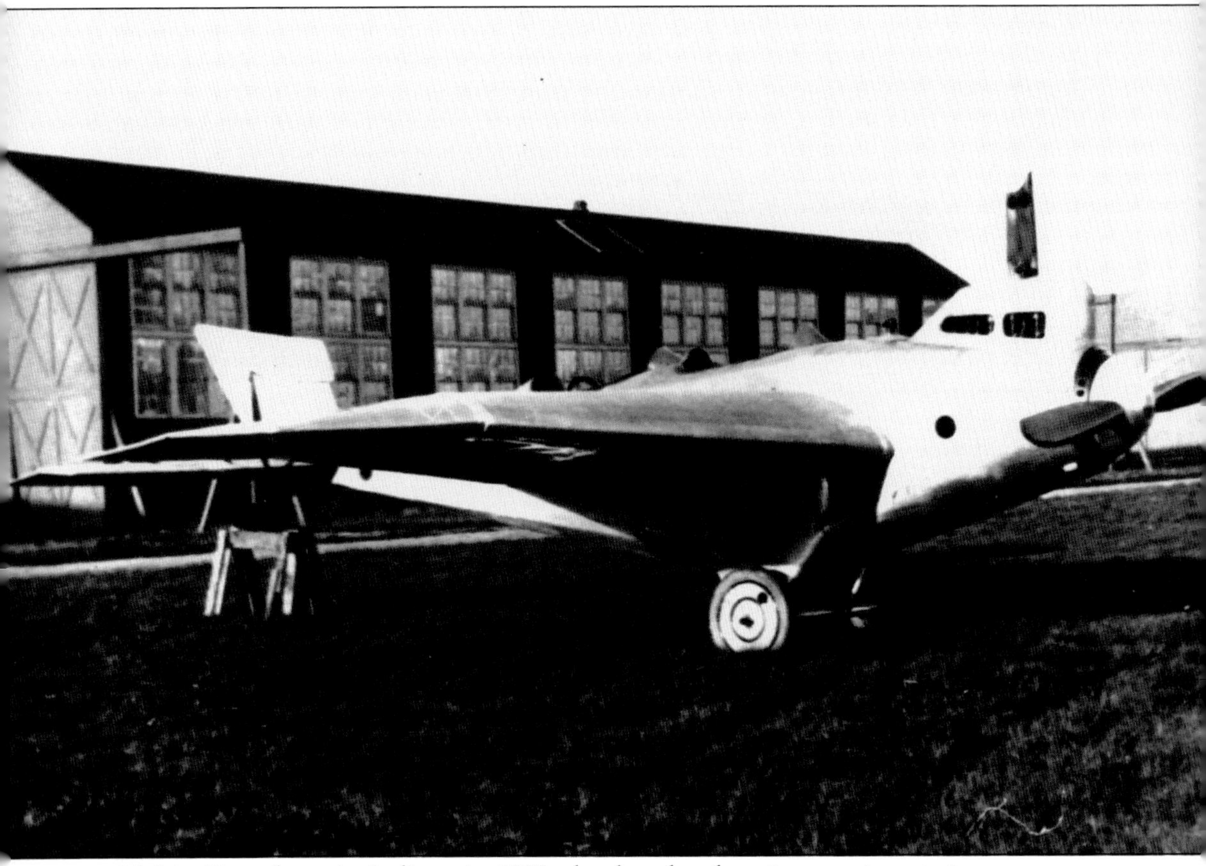

Melvin Dare, born in Indiana in 1888, developed a deep interest in aviation as a teenager. Dare felt that he could improve the performance of an airplane by observing the flight of birds. He went to Texas to study vultures and noted how they varied the curvature of their wings in flight. He received a patent for a Variable Camber Wing and moved to Detroit to seek financing to build an airplane incorporating his ideas. In 1922, he built an all-metal monoplane with variable camber wings, originally powered by a 140-horsepower Benz (seen here) and later by a 400-horsepower Liberty engine. Eddie Stinson test flew the airplane at Selfridge Field in Mount Clemens with questionable results. In 1925, Dare built a second airplane for the Guggenheim Safe Airplane Competition. It was a two-seat cabin monoplane with an improved camber actuating mechanism The first flight was made by Slygh Brown at Scully Field, Detroit, in late 1928. During a demonstration flight on July 17, 1929, one wing broke off and both Dare and Brown were killed.

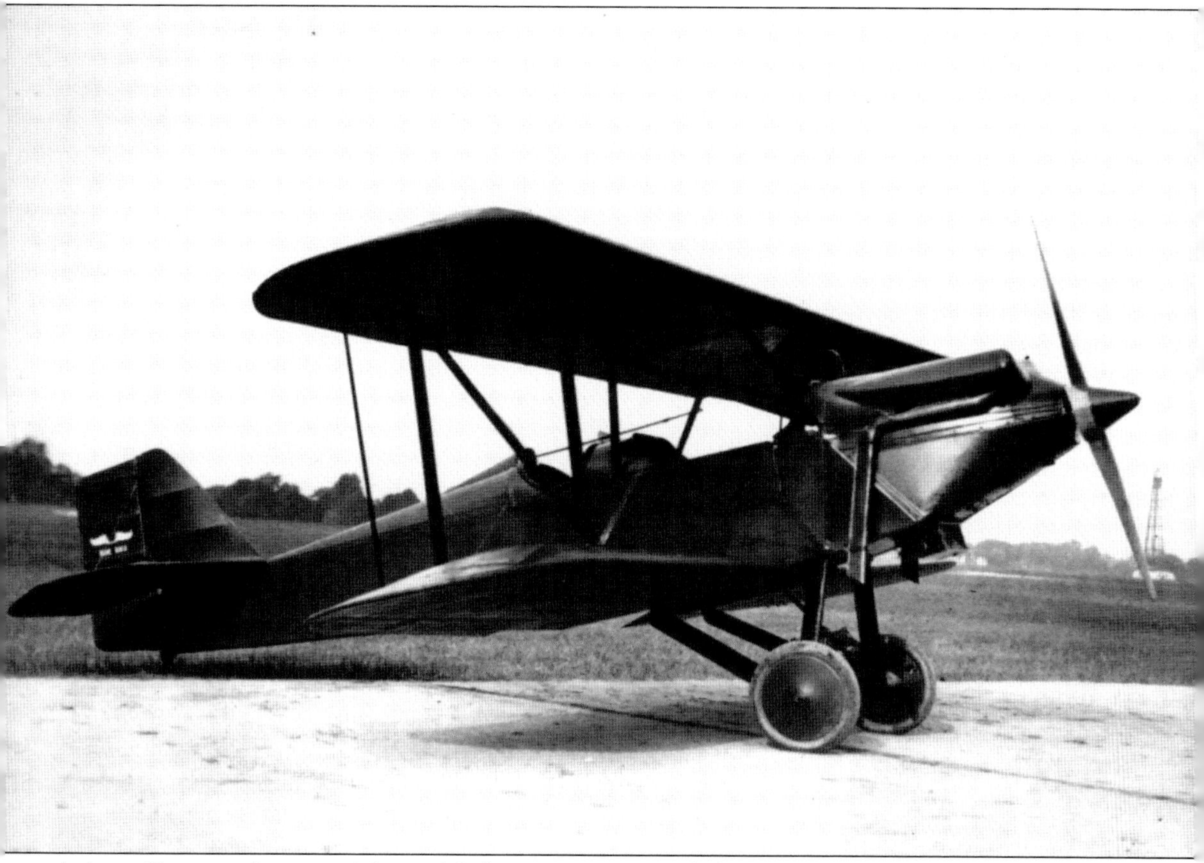

Aubrey Hess was born in Adrian, Michigan, in 1899 and during World War I served in the U.S. Army Air Service as an aeronautical engineer. He returned to civilian life in 1925 and built a three-place biplane powered by a 180-horsepower Hispano-Suiza engine. Hess painted the airplane dark blue and named it the Blue Bird. First flown in mid-1925, he and his brother Adrian formed the Hess Airplane Company, Wyandotte, Michigan, to produce the airplane. Blue Birds were sold in kit form as well as complete airplanes, so it is difficult to determine how many were actually built. Some sources show 6, others as many as 20. One participated in the 1926 Ford Reliability Air Tour (seen here) and another in the Dole Race. In 1927, Aubrey met with officials from Alliance, Ohio, who wanted to attract new business to their community. He accepted their offer and in 1928 moved there. On October 12, 1929, Aubrey was killed testing an Alliance Argo biplane.

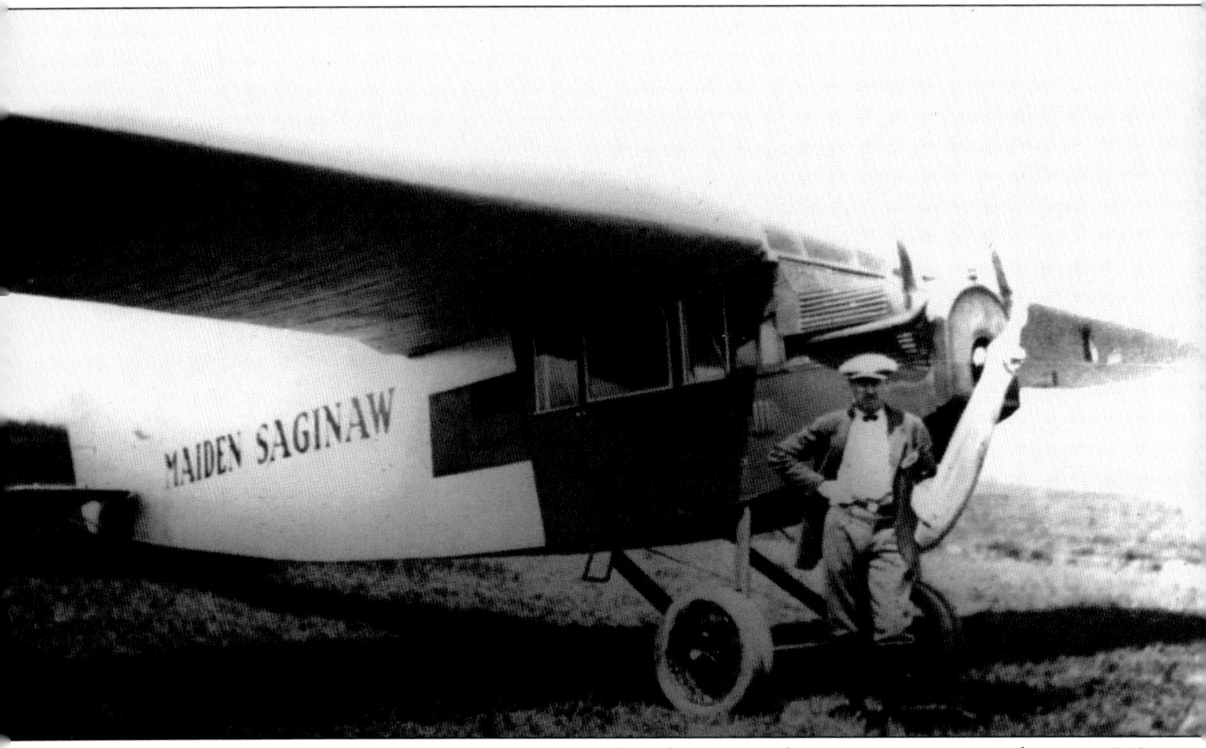

Born in Illinois in 1896, Walter Carr wanted to become a locomotive engineer but in 1912 saw an airplane and decided to become an aviator. He went to Chicago in 1914 to take flying lessons and soloed that June. Shortly after soloing, he was hired by the Humbolt Exhibition Company as an exhibition pilot. During World War I, he served as a flight instructor and after the war barnstormed around the Midwest. In 1924, he settled in Saginaw where he operated a flying service. Carr saw a need for a cabin-type airplane, so with financial help from the Savage brothers and John Coryell, he formed the CSC Aircraft Company. Carr built this three-place monoplane named *Maiden Saginaw* and test flew it in May 1925. With a 100-horsepower OXX-6 engine, the airplane was underpowered and did not perform well. Carr's partners lost interest in the airplane, and it was later scrapped. (Courtesy William Clunie collection.)

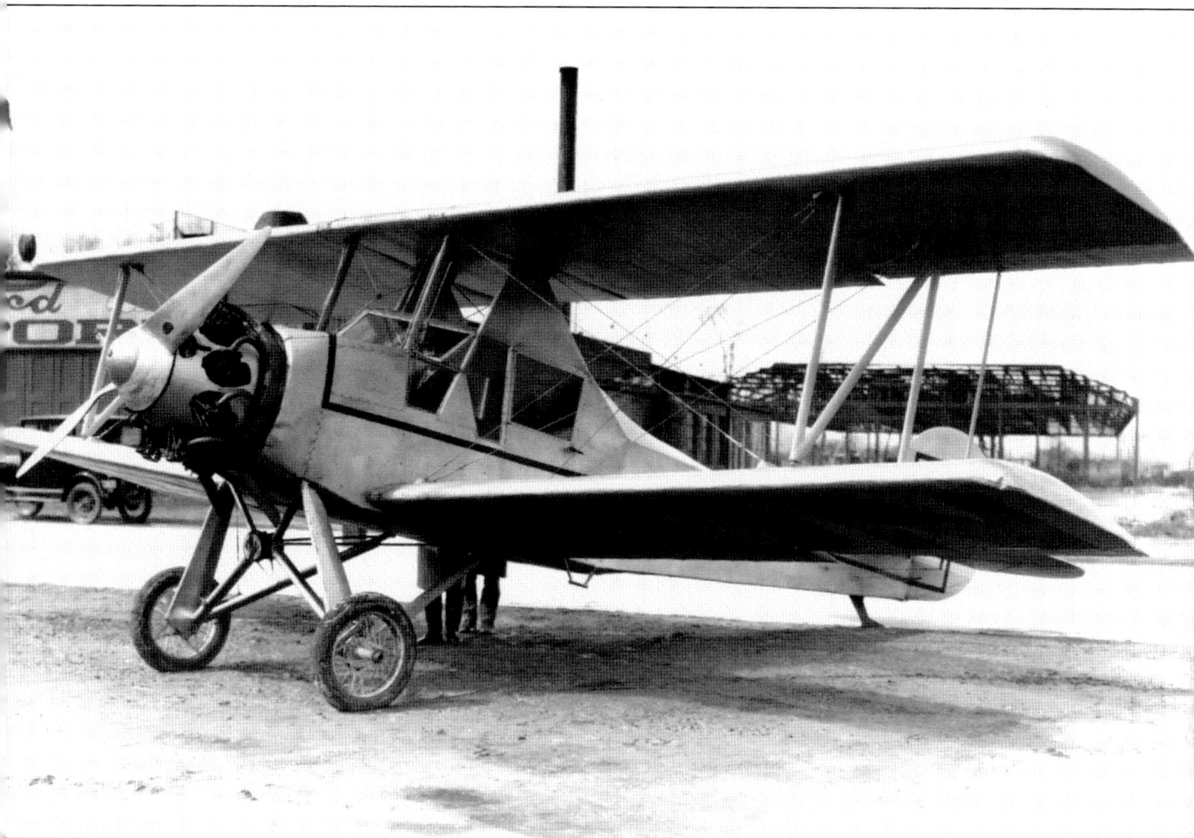

Edward Stinson was born in Alabama in 1894 and learned to fly in 1915. In 1917, the Stinson family, including Eddie's older sisters Katherine and Marjorie, both early exhibition flyers, opened a flying school in San Antonio, Texas. Eddie's years as an instructor and barnstormer honed his flying skills, and in 1922, he moved to Detroit to operate a charter flying service. In the fall of 1925, Eddie presented his ideas for a cabin-type airplane to the Detroit Board of Commerce and with their support formed the Stinson Airplane Syndicate. In a warehouse on Congress Street, Eddie built this four-place biplane featuring a self-starter, wheel brakes, and a heated cabin. Eddie test flew the SB-1 Detroiter at Selfridge Field in Mount Clemens on January 25, 1926. The airplane attracted much interest, and Stinson had received many orders. In May 1926, with financial assistance from William Mara, the Stinson Aircraft Corporation was formed and moved to Northville, and by the end of 1926, 10 SB-1s had been sold.

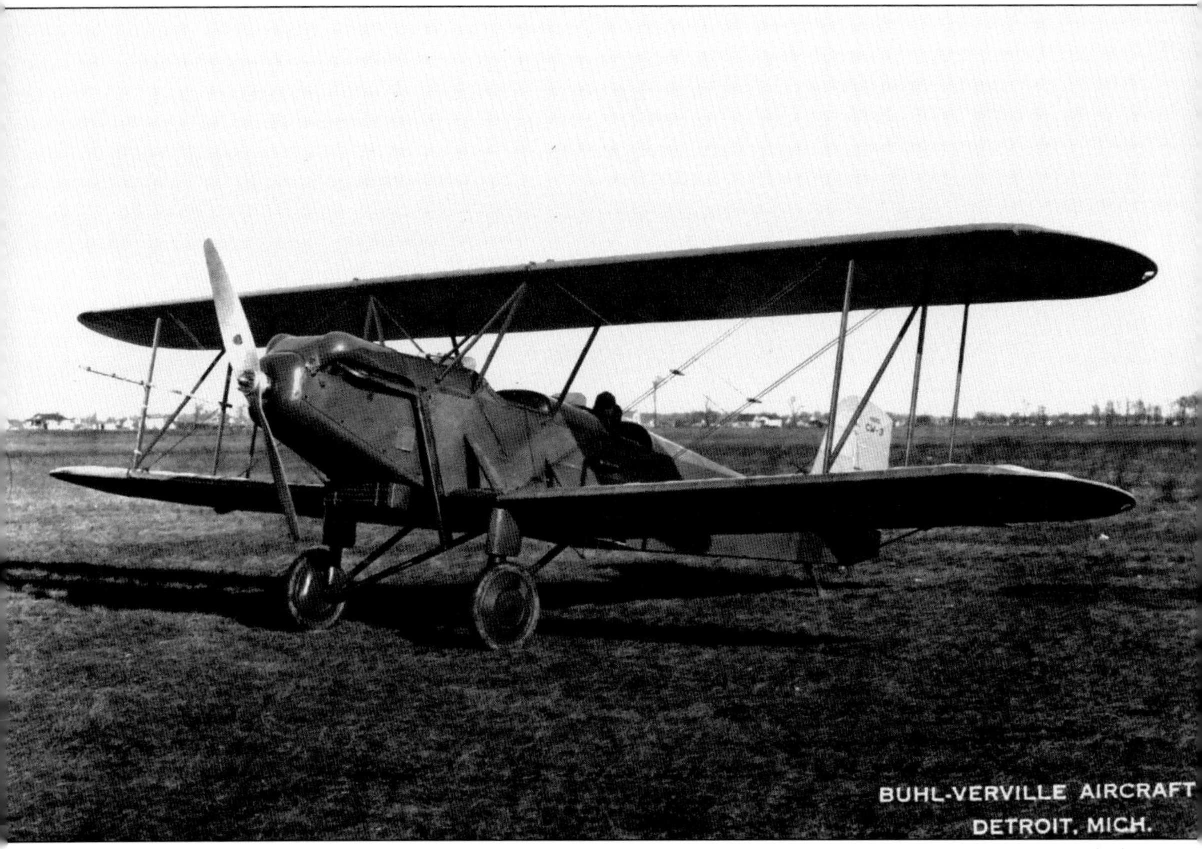

After Fred Verville closed down his General Aeroplane Company in 1916, he went to work for Fisher Body in Detroit to work on the Liberty Plane program. When the war ended, Verville was hired by the U.S. Army Air Service, Dayton, where he designed the VCP-1 pursuit, the R-3 racing plane, and the Sperry Messenger. But Verville was more interested in building commercial aircraft, so in March 1925, he moved to Detroit and joined forces with Lawrence Buhl to form the Buhl-Verville Aircraft Corporation. Buhl was a prominent Detroit industrialist who wanted to enter the growing field of aircraft manufacturing, and Verville helped him get started. In a corner of the Buhl Stamping Plant in Detroit, Verville designed the CW-3 Airster, a three-place biplane powered by an OX-5 engine and with folding wings to reduce storage space. The prototype was test flown by Jack Hunt in December 1925 at Packard Field in Detroit. Three Airsters were built in the Buhl Stamping Plant before Larry Buhl moved the company to Marysville, Michigan.

Four

Government Regulations Introduced 1927–1929

In the history of aviation in America, 1927 was a landmark year. Pres. Calvin Coolidge had signed into law the Air Commerce Act, to be enforced by the Aeronautics Branch of the U.S. Department of Commerce and to become effective on January 1, 1927. The new law had broad powers covering the licensing of pilots, mechanics, and aircraft and required that all aircraft be identified with a license number, known as an "N" number.

Furthermore the regulations mandated that all new commercial aircraft meet stringent design requirements supported by engineering data and flight tests. After complying with those requirements, the manufacturer would be issued an approved type certificate (ATC) indicating that the airplane had met federal standards. Since the earliest days of aviation, pilots and aircraft manufacturers had been operating without rules or regulations and most Americans viewed flying as unsafe and dangerous. The goal of the new law was to assure the public that flying was a safe form of transportation and also to promote the growth and development of commercial aviation.

In May 1927, an event took place that was to have a profound impact on the future of the American aircraft industry. Charles Lindbergh, born in Detroit in 1902, flew solo from New York to Paris in a single engine monoplane, the *Spirit of St. Louis*, and America became air-minded overnight. One of Lindbergh's greatest contributions to aviation was his nationwide tour following that historic transatlantic flight. Between June and October 1927, Lindbergh, flying his famous Ryan-built monoplane, visited 82 cities in all 48 states, on time and on schedule. He gave speeches promoting aviation to thousands of spectators creating what came to be known as the "Lindbergh Boom."

America's aircraft industry expanded rapidly following Lindbergh's flight as new aircraft companies were formed to cash-in on a booming demand for anything associated with aviation. Hundreds of aircraft manufacturers opened for business with new designs, but it all came to an abrupt end when the stock market crashed in October 1929.

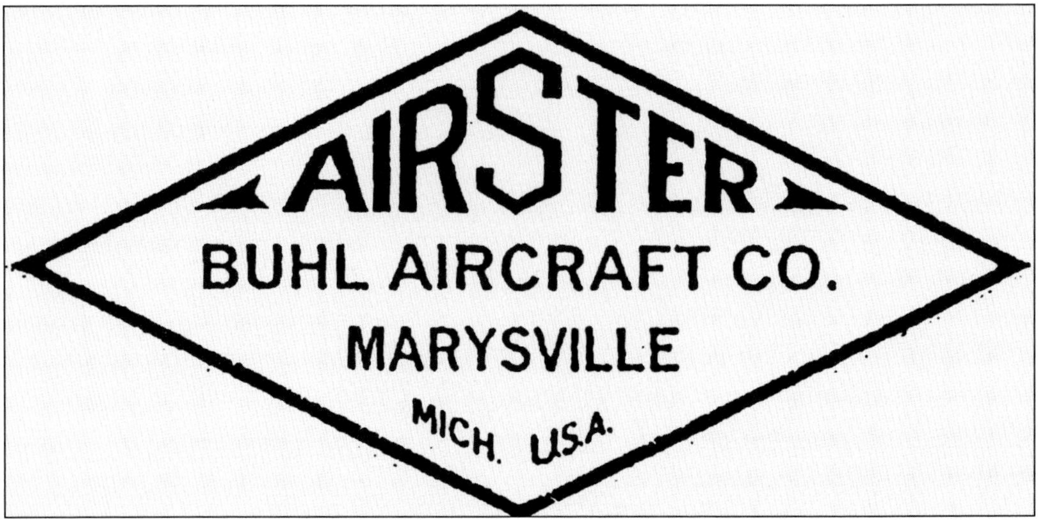

This diamond-shaped Buhl Aircraft Company emblem was painted on the fin of Buhl Airsters and Airsedans that were built in Marysville, Michigan, by the Buhl Aircraft Company. The Buhl-Verville Airsters that were built in Detroit in 1925 and 1926 used a similar diamond-shaped logo but with the city of "Detroit" instead of "Marysville."

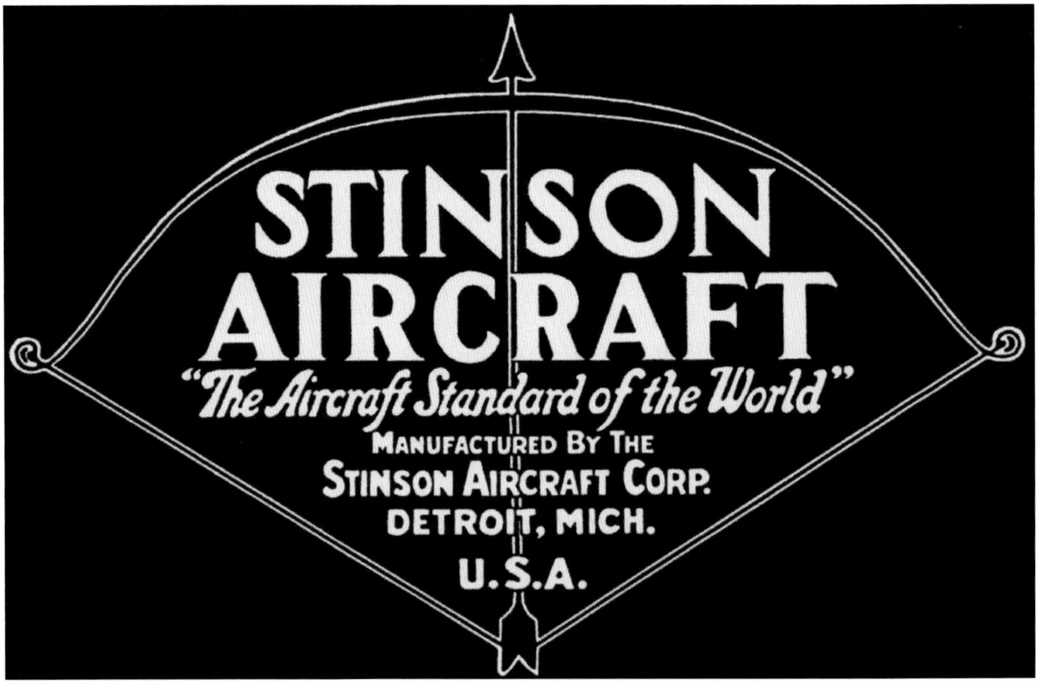

This Stinson emblem was painted on the fin of Stinson airplanes and varied in size, shape, and design over the years. Eddie Stinson was proud of his Cherokee Indian heritage and this bow and arrow logo reflected his Native American roots.

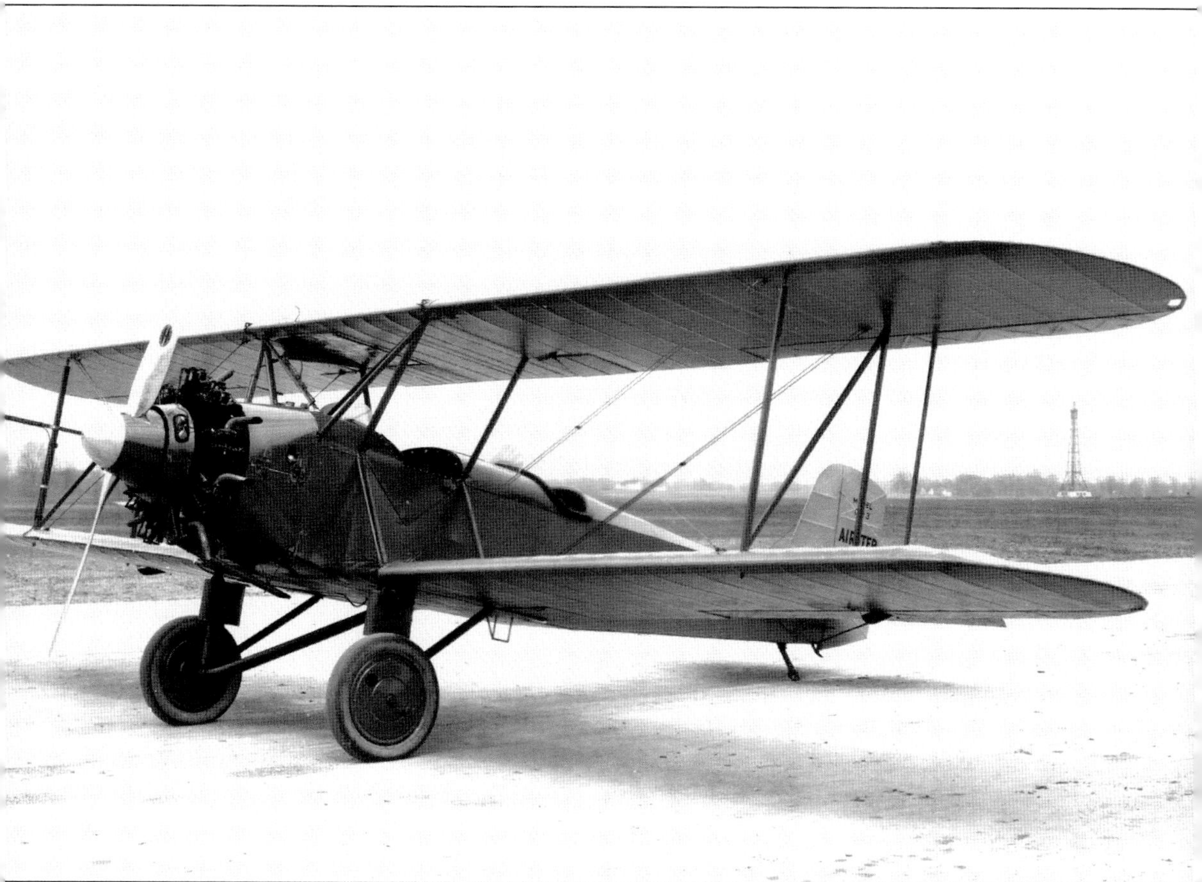

Fred Verville, chief engineer for the Buhl-Verville Aircraft Company, was planning to build an Airster with a 200-horsepower Wright J-4 engine. However, when Larry Buhl decided to move the company to Marysville, Michigan, Verville resigned, and Etienne Dormoy was hired as his replacement. Dormoy, born in France in 1885, had come to America during World War I to work with U.S. Army Air Service engineers in Dayton, where he first met Verville. It was Verville who had recommended that Buhl hire Dormoy as the new chief engineer. Dormoy's first assignment after the move was to get the OX-5-powered Airster back into production. In addition, Dormoy also built a J-4-powered Airster as Verville was planning to do and applied for an ATC. ATC No. 1 was awarded to the J-4 Airster in March 1927, the first airplane to receive an airworthiness certificate from the U.S. Department of Commerce. A total of 12 Airsters were built, 3 in Detroit and the rest in Marysville, as CW-3 (OX-5 engine), CA-3 (Wright J-4 engine), CA-3A (Wright J-5 engine), plus one military trainer version.

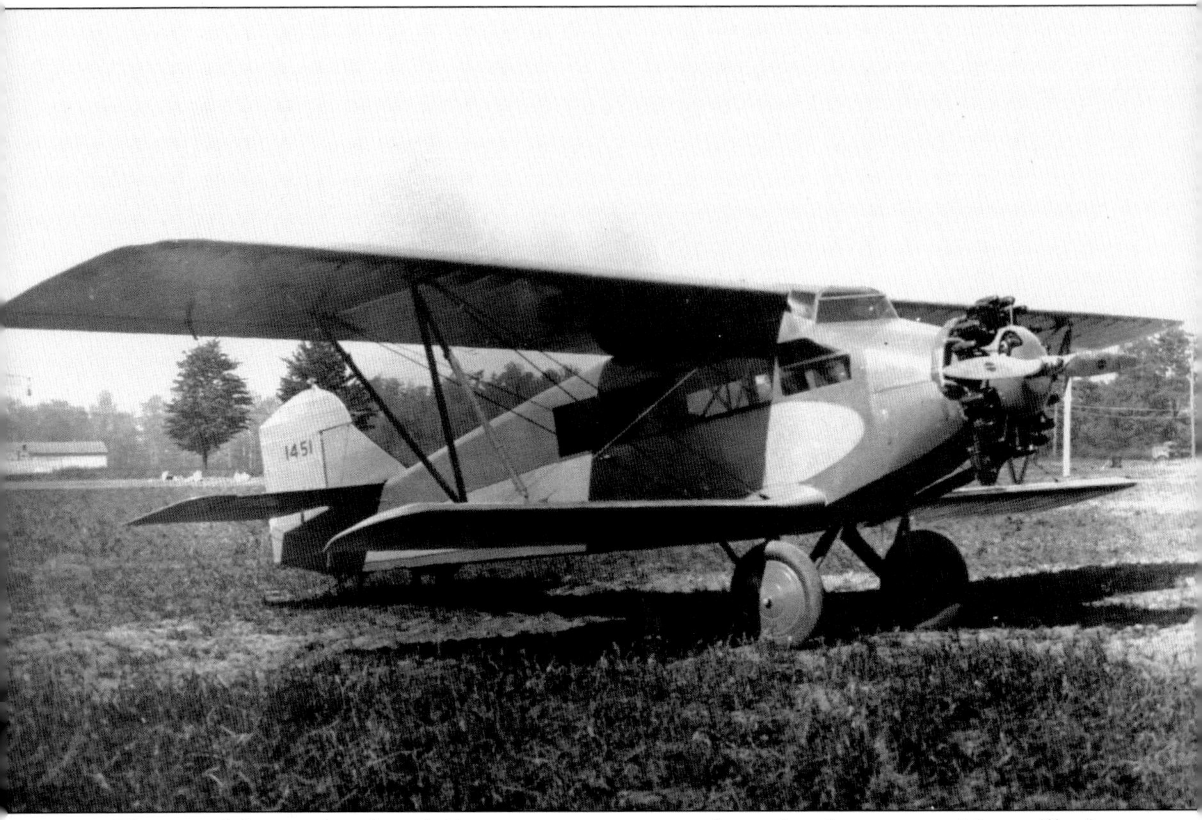

The era of the open-cockpit biplane was nearing an end, so after the move to Marysville, Larry Buhl had Etienne Dormoy design this five-place cabin biplane powered by a Wright J-5 engine. It featured a roomy cabin for four with the pilot up front in an elevated seating position. The Buhl CA-5 Airsedan was awarded type certificate No. 12 in September 1927. Many were sold for business use, and one CA-5 finished 10th in the 1927 Ford Reliability Air Tour. In 1928, Buhl introduced the CA-5A with a more luxurious interior and other refinements. One CA-5A, *The Angelino*, set an endurance record of over 246 hours aloft using midair refueling. The basic CA-5 sold for $12,500, and a total of 15 were built.

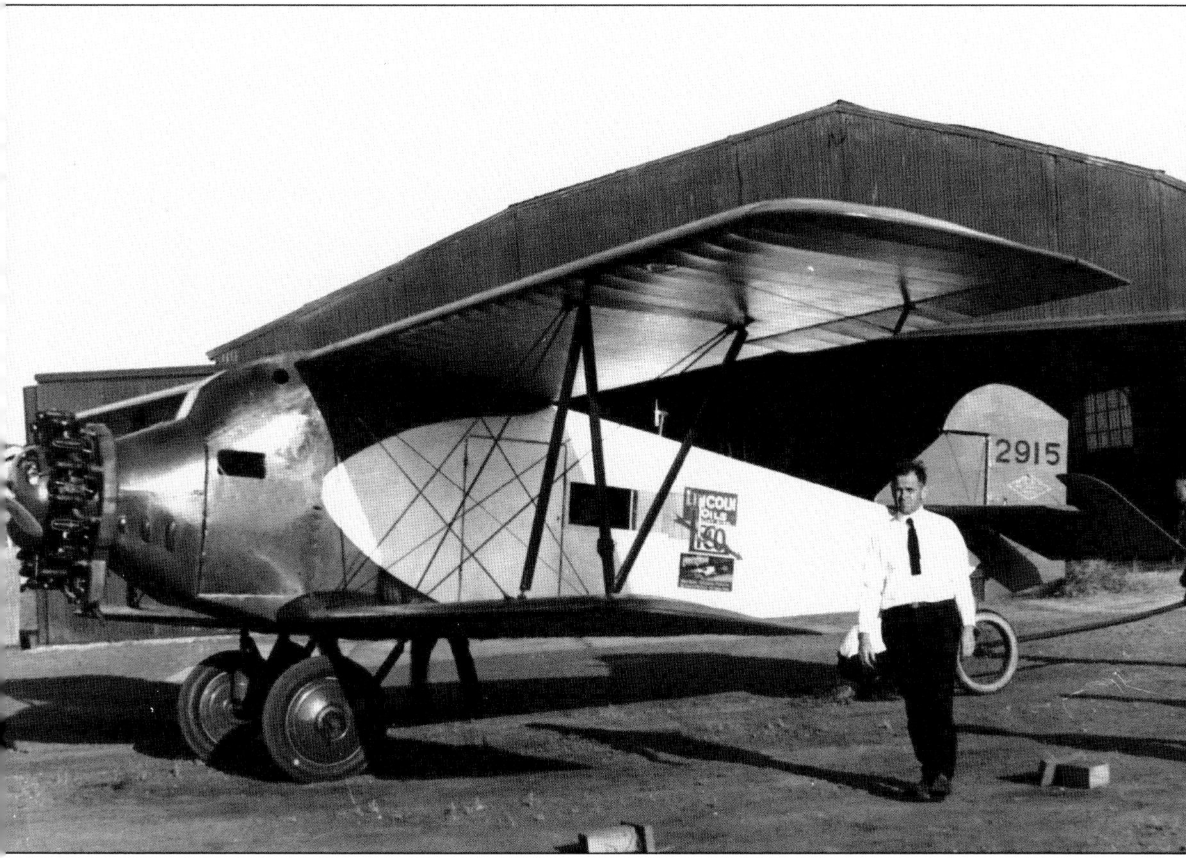

Perhaps the best known Buhl CA-5 Airsedan was built for William Maloska, owner of a chain of service stations in Flint. When the Dole Race was announced offering $25,000 to the first airplane to fly from California to Hawaii, Maloska felt that his Airsedan could easily win the prize money. Preparations included increasing the area of the lower wing and installing a 400-gallon fuel tank. Maloska's pilot was John "Augie" Pedlar with Vilas Knope as navigator. Pedlar also invited 22-year-old Mildred Doran to accompany him on the trip, and the Airsedan was named *Miss Doran* in her honor (seen here, with Pedlar at right). On August 16, 1927, nine airplanes were lined-up at the Oakland, California, airport for the race. Five crashed attempting to takeoff with heavy fuel loads, and only two airplanes reached Hawaii. Winners Art Goebel and Bill Davis took 26 hours and 19 minutes to complete the trip. *Miss Doran* and another airplane were lost at sea, and two other airplanes were lost searching for the missing flyers.

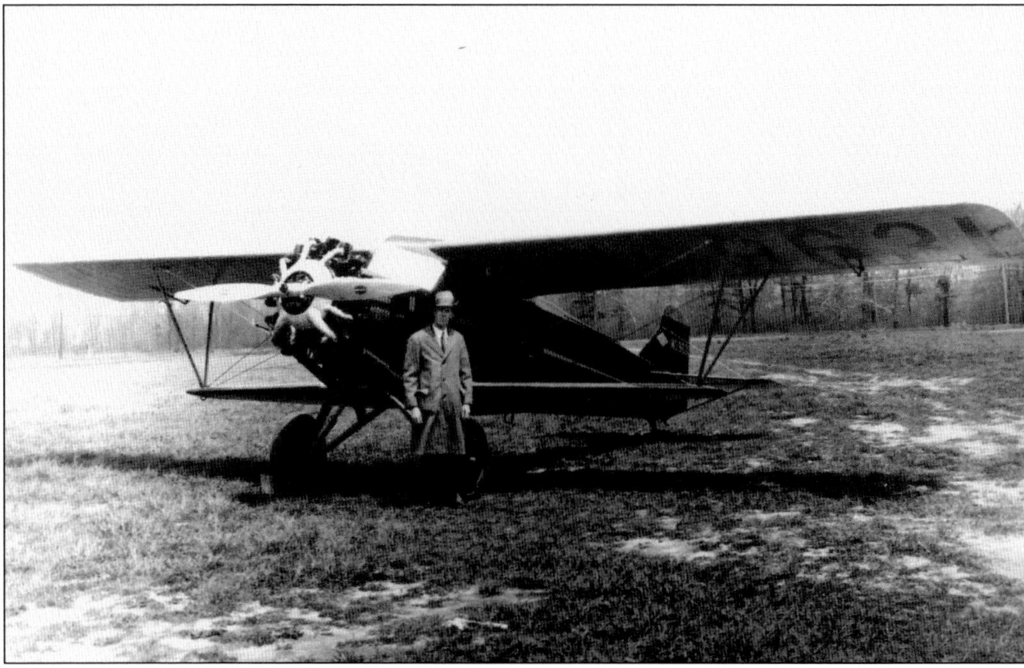

In May 1928, the Buhl Aircraft Company introduced the CA-3C Sport Airsedan (seen above, with engineer Etienne Dormoy), a smaller, three-place version of the CA-5. Its most distinctive feature was the small size of the lower wing, which had less than half the area of the upper wing, a configuration known as a sesquiplane. It was powered by either a Wright J-5 or J-6 engine. A variant, the Junior Airsedan, used a 110-horsepower, Detroit-built, Warner Scarab engine. Some 13 Sport and Junior Airsedans were built. Another Dormoy design, introduced in late 1928, was the CA-8A Senior Airsedan (seen below), an eight-place sesquiplane powered by a 300-horsepower Wright J-6 engine. Four were built before the Great Depression ended a demand for large airplanes.

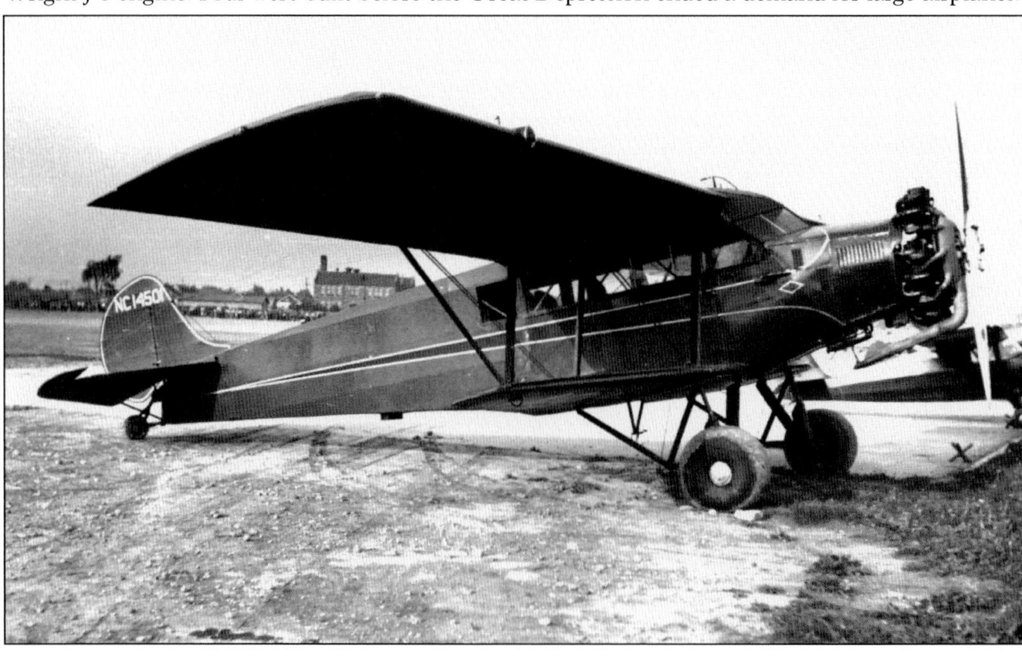

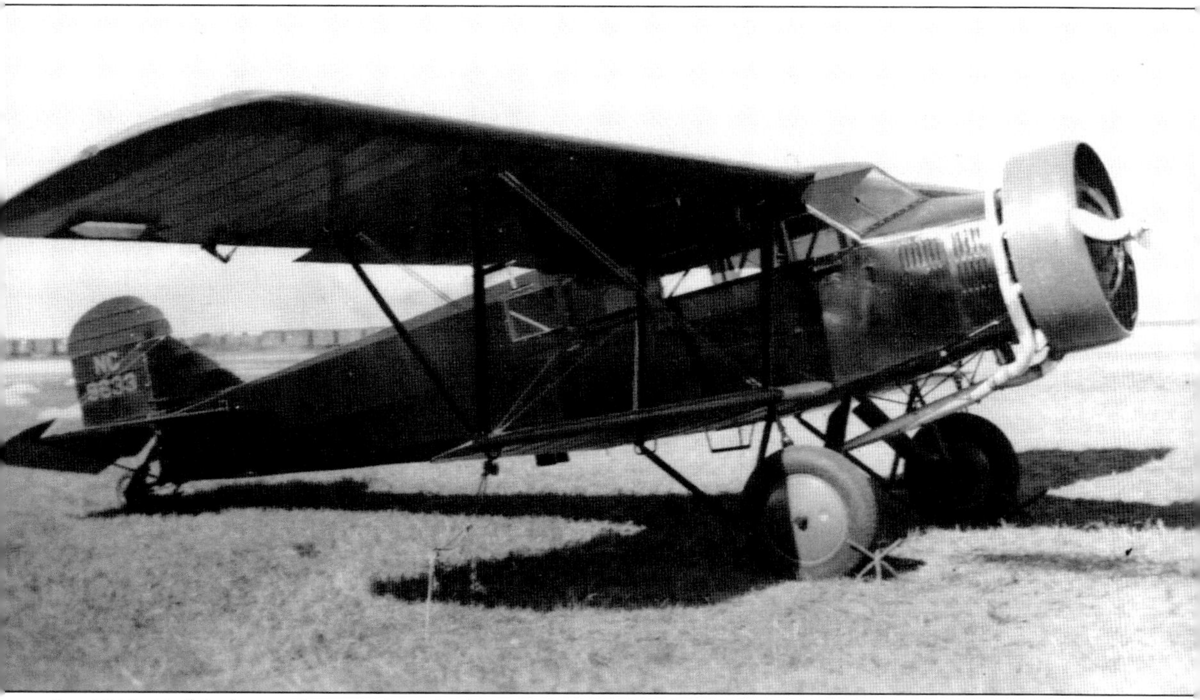

In 1929, Buhl Aircraft introduced the CA-6 Standard Airsedan, which became the most popular in the series. A sesquiplane, it had a slimmer fuselage than earlier models and was a cross between the small CA-3C and the larger CA-8A. Most of the Standard Airsedans were powered by a 300-horsepower Wright J-6 engine (the Wright J-5 Whirlwind that had carried Lindbergh and the *Spirit of St. Louis*, safely across the Atlantic had been discontinued in 1928 and its replacement was the Wright J-6 series). The CA-6 was also available with a 450-horsepower Pratt and Whitney Wasp. Some 16 CA-6s were built and widely used for passenger and cargo operations in the Pacific Northwest and in Canada.

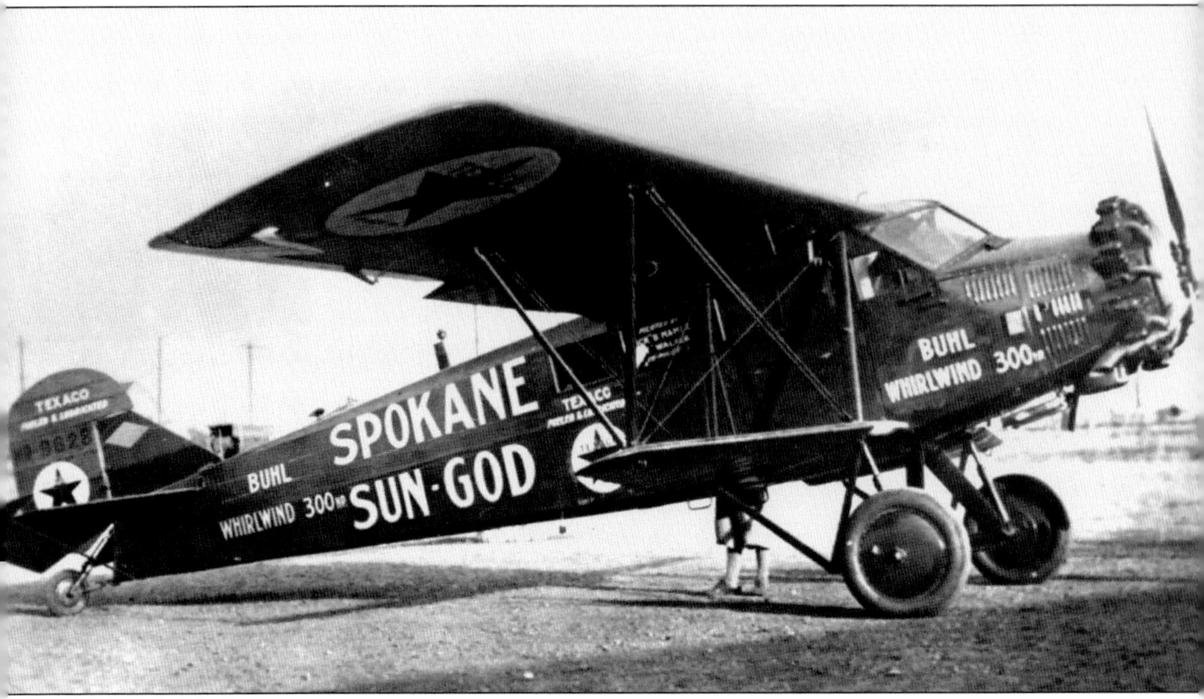

A memorable long-distance flight was made in this Buhl Airsedan, the *Spokane Sun-God*, a CA-6 modified to carry 320 gallons of fuel and with an opening on top of the fuselage for midair refueling. The *Spokane Sun-God* established a record by flying nonstop from Spokane, Washington, to New York City and return. Refueling took place at scheduled rendezvous points along the route. Pilots Nick Mamer and Art Walker took-off from Spokane's Felts Field on August 15, 1929. With no radios and unreliable weather forecasting, it was a hazardous undertaking. On one occasion, the refueling hose was severed by the *Spokane Sun-God*'s propeller, spraying fuel over both airplanes, and thunderstorms and smoke from forest fires made it difficult to locate the tanker planes. Despite many such obstacles, the *Spokane Sun-God* returned safely to Spokane after 120 hours and one minute in the air.

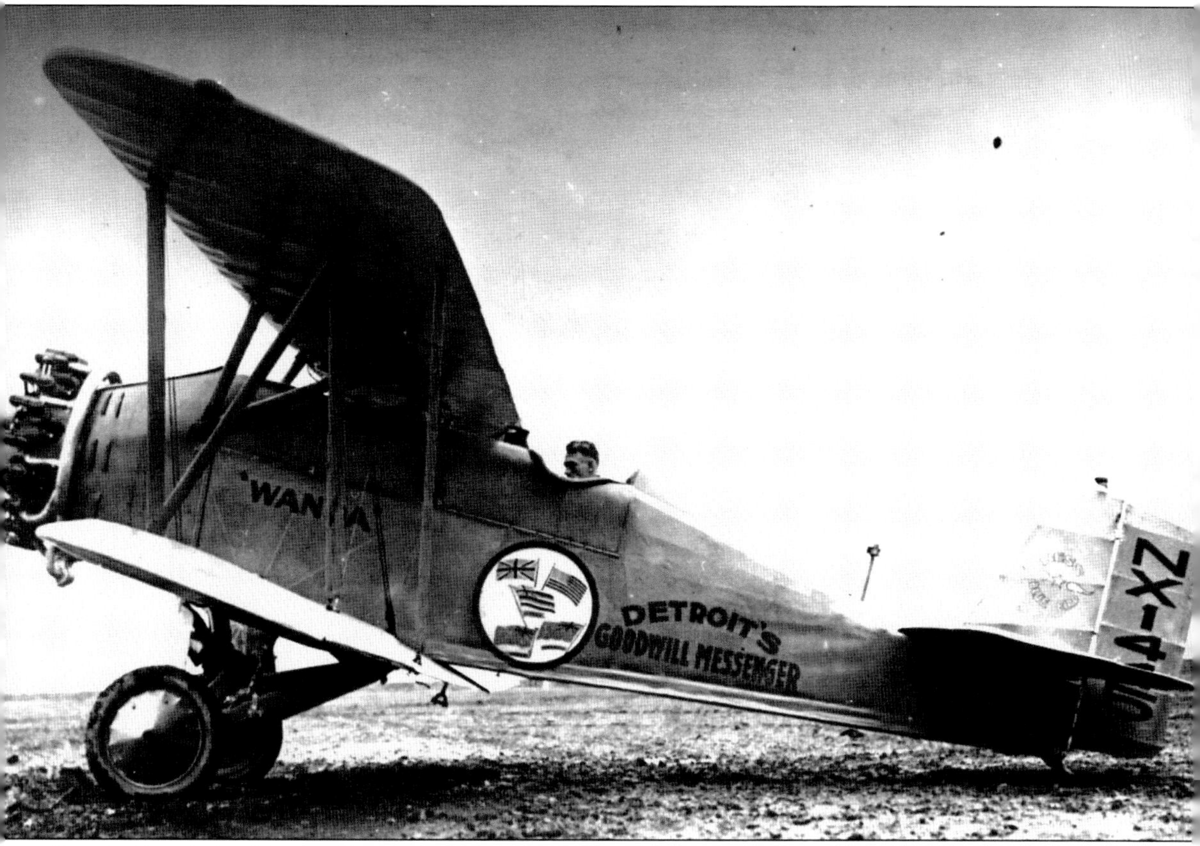

British Royal Air Force pilot Capt. Frederick Giles, born in 1898, moved to Australia to fly air mail routes. In 1927, he decided to enter the Dole Transpacific Air Race. While sailing to America, Giles met William Rosewarne, an Australian then living in Detroit, who offered to sponsor Giles's flight. They ordered a Hess Blue Bird, built in Wyandotte, Michigan, with a Wright J-5 engine and extra fuel tanks. Giles planned to fly nonstop from Detroit to California, enter the Dole Race to Hawaii, then continue on to New Zealand. The airplane was named *Miss Wanda* after Aubrey Hess's sister. Giles left Ford Airport on August 19, 1927, three days after the Dole Race deadline. With several mishaps en route, he did not reach California until October. Giles took off for Hawaii on November 10 but returned after encountering thick fog. Some 12 days later he tried again, and again returned, crash landing near San Simeon, California. Giles was unhurt, and the Blue Bird was later rebuilt.

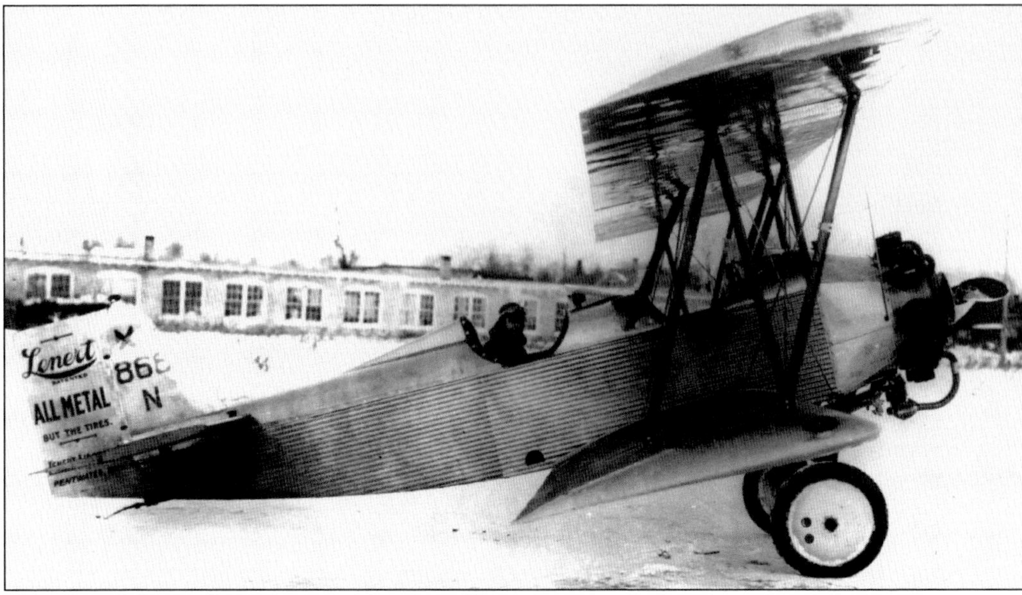

In 1925, Willy Lenert arrived in Pentwater, Michigan, to establish an airplane factory. He held several patents for all-metal airplanes, sold stock to local residents, and formed the Lenert Aircraft Company. His first design was an all-metal biplane with an OX-5 engine. A second airplane used a Continental A-70 engine (seen here). Lenert later moved to Muskegon, Michigan, promoting the airplane with the slogan "All metal but the tires."

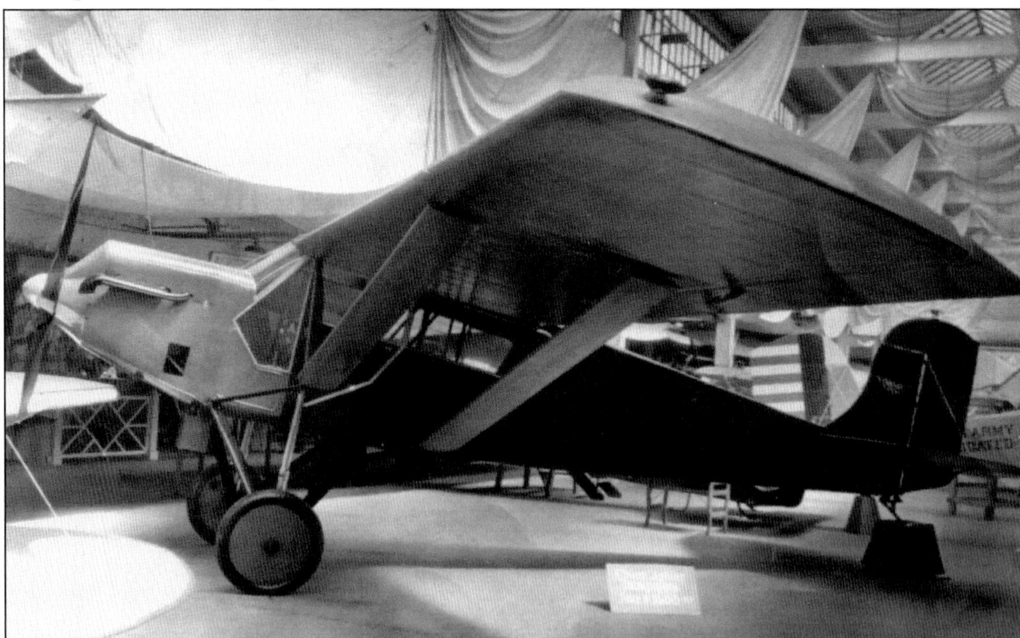

Frank Overcashier was born in Pennsylvania in 1887, and his family later moved to a farm near Quanicassee, Michigan. In 1922, he bought a surplus Curtiss Jenny, taught himself to fly, then opened a flying school in Highland Park, a Detroit suburb. In 1928, he formed the Overcashier Aircraft Company and built three Overcashier O-12 cabin monoplanes powered by OX-5 engines. Overcashier died in 1948.

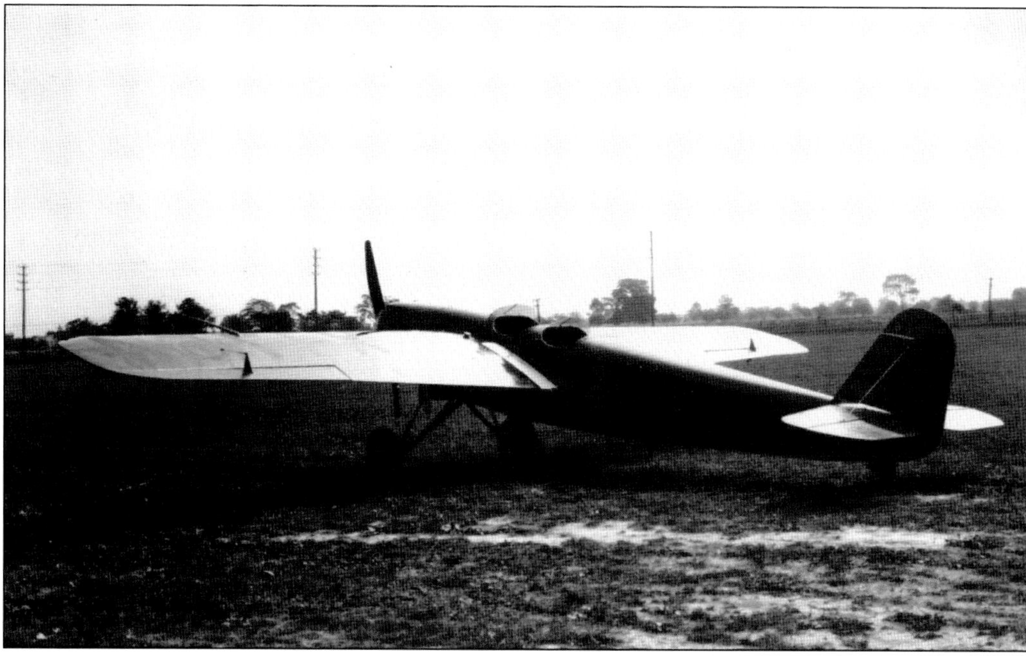

The W. F. Stewart Company was established in Flint in 1908 to build wooden automobile bodies. In 1927, president Sidney Stewart decided to manufacture airplanes and hired Jack Hunt and Lionel Kitchen to design this all-wood, four-place monoplane The occupants sat side-by-side in two tandem and open cockpits. The OX-5 powered Stewart X-1 had a varnished mahogany finish and first flew in 1928.

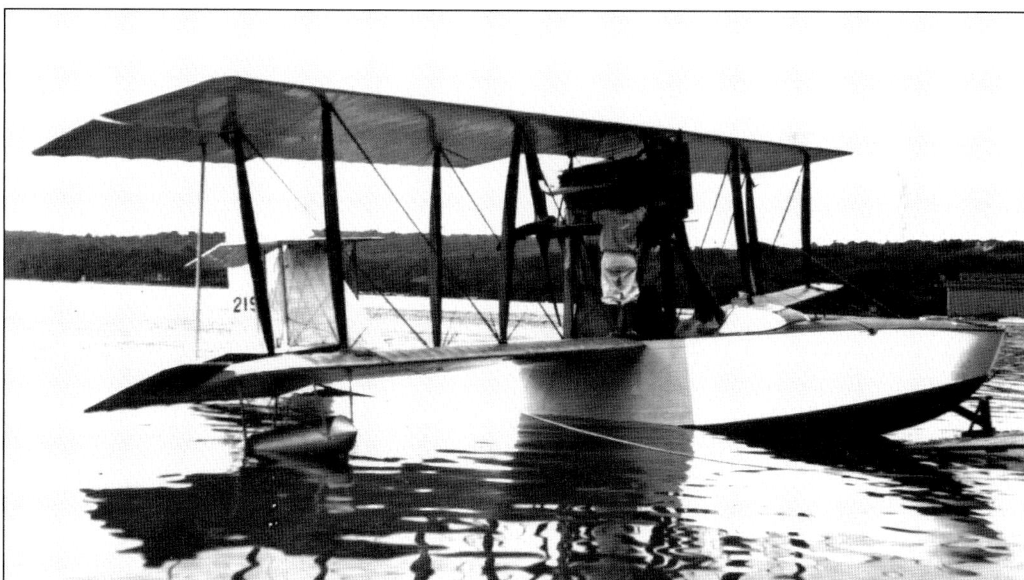

Ross Judson, born in Minnesota in 1881, was cofounder of the Autocar Company in Chicago. In 1905, Autocar, renamed Continental Motors, moved to Muskegon to manufacture automobile and aircraft engines. In 1922, Judson had a flying boat built in the Continental Motors factory for his own personal use. The Kanter-Judson *Air Yacht*, originally a triplane, was later converted to a biplane, as shown here. (Courtesy Dave Ostrowski collection.)

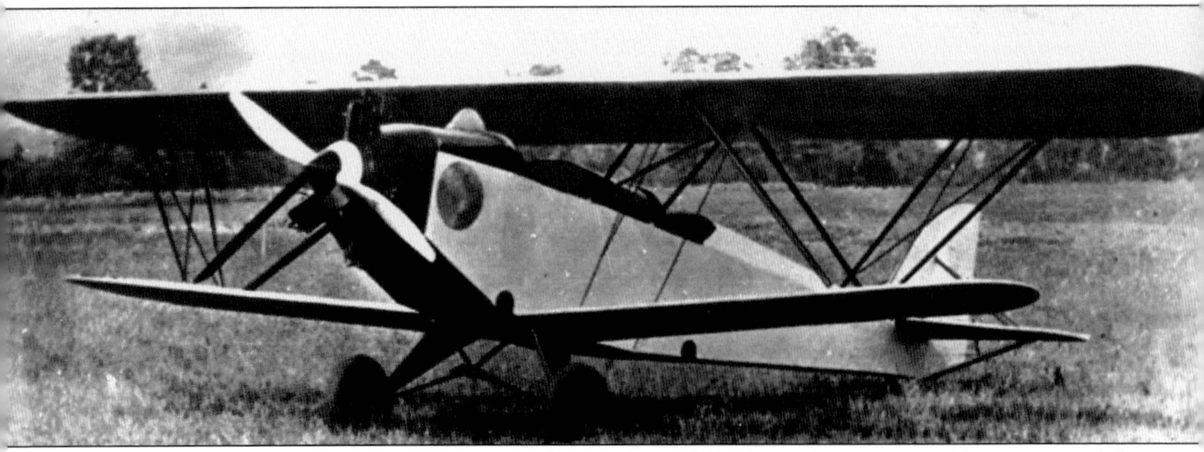

Ivan Driggs was born in Lansing in 1894 and attended Michigan State College. In 1915, he moved to Dayton, Ohio, to work for the U.S. Army Air Service as an aircraft designer. He left the U.S. Army Air Service in 1924 to join the Johnson Aircraft Company in Dayton, where he designed the Driggs-Johnson DJ-1, a single-place monoplane that won the Dayton News Trophy Race. Driggs left the Johnson Aircraft Company to form his own company and built three improved versions of the DJ-1, known as Driggs Darts. In 1927, Driggs moved back to Lansing to form the Driggs Aircraft Company with backing from local businessmen. The company built the Driggs Dart II (seen here), a two-place sesquiplane with Warren-truss interplane struts and powered by a 36-horsepower Anzani engine. Some 15 were built before the U.S. Department of Commerce canceled the ATC (No. 15) because of unacceptable spin characteristics.

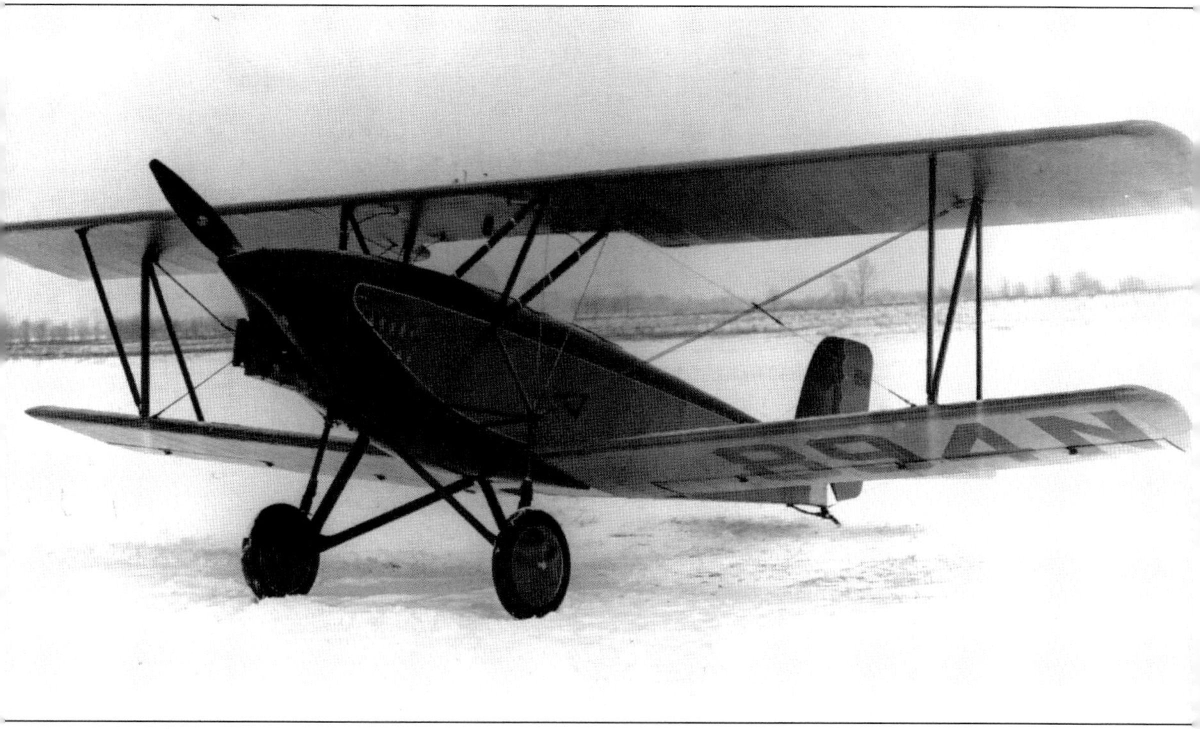

In December 1928, Driggs Aircraft introduced the Skylark at the Chicago Aero Show, a two-place biplane with an eye-catching red, black, and yellow color-scheme. The power plant was a 75-horsepower, four-cylinder, air-cooled Rover engine designed by Harold Morehouse and built in Lansing. The Skylark sold for $3,485, but many of the orders taken at the Chicago show were canceled because of the depressed economy. Driggs built 17 Skylarks before closing down in 1931. In December 1933, the manufacturing rights were sold to Skylark Aircraft Company in Muskegon, and they assembled one from existing parts. The manufacturing rights were later sold to Phillips Aviation in California. Driggs left Lansing in 1933 to work for Consolidated, Monocoupe and Luscombe. In 1941, he was named director of research for the Bureau of Aeronautics, Washington, D.C. Ivan Howard Driggs died in 1955.

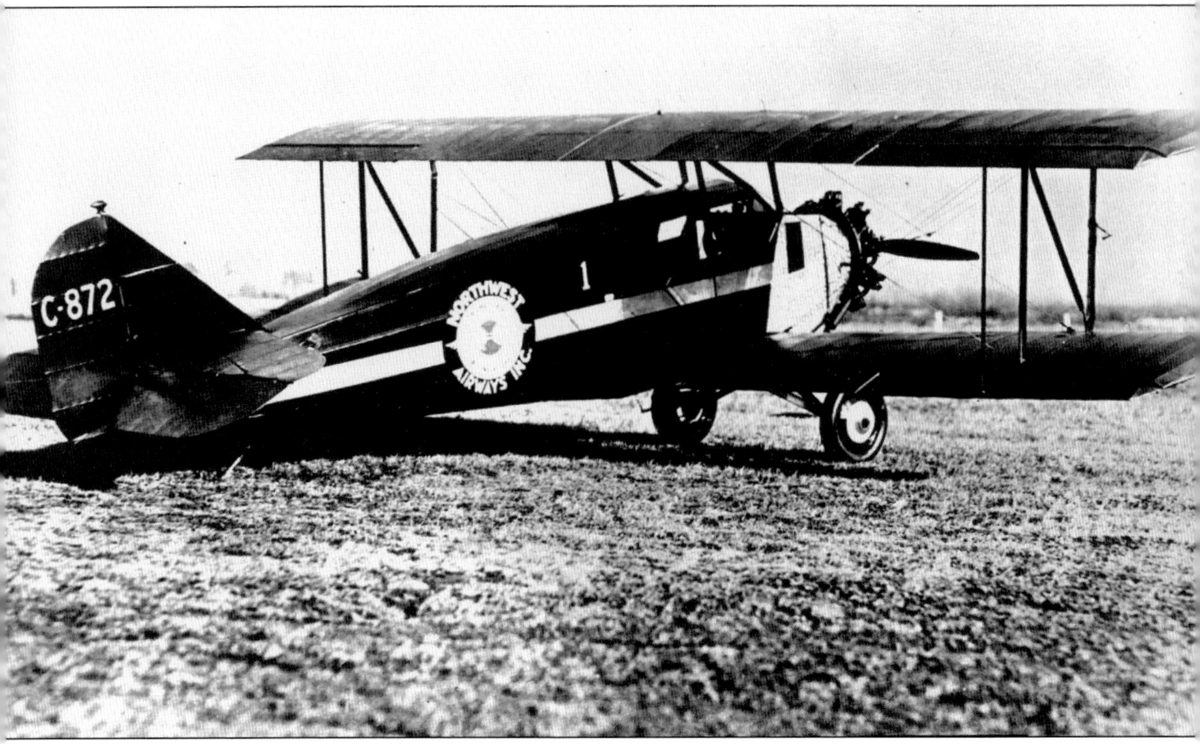

After moving to Northville, Michigan, in 1926, Stinson Aircraft began building a production version of the SB-1 Detroiter. Now powered by a 220-horsepower Wright J-5 engine, it had seating for five, a revised vertical tail, and other changes. Vice president William Mara, a super salesman, convinced many business executives they needed an airplane to be successful. Horace Dodge and cousin John, both of the automobile company, bought Detroiters for business and pleasure. Others were sold to Florida Airways, Wien Alaska Airways, and to Detroiter Edward Schlee, owner of Wayne Oil company. In August 1926, Eddie Stinson, flying Schlee's Detroiter, finished third in the Ford Reliability Air Tour. In September 1926, Bill Mara arranged a meeting between Col. L. H. Brittin, from Minnesota, and several prominent Detroit businessmen who agreed to help Brittin finance a new airline. The outcome was Northwest Airways and an order for three Detroiters, the humble beginning of today's Northwest Airlines. Stinson built 26 SB-1 biplanes at Northville until it was replaced by the new SM-1 monoplane. (Courtesy Northwest Airlines.)

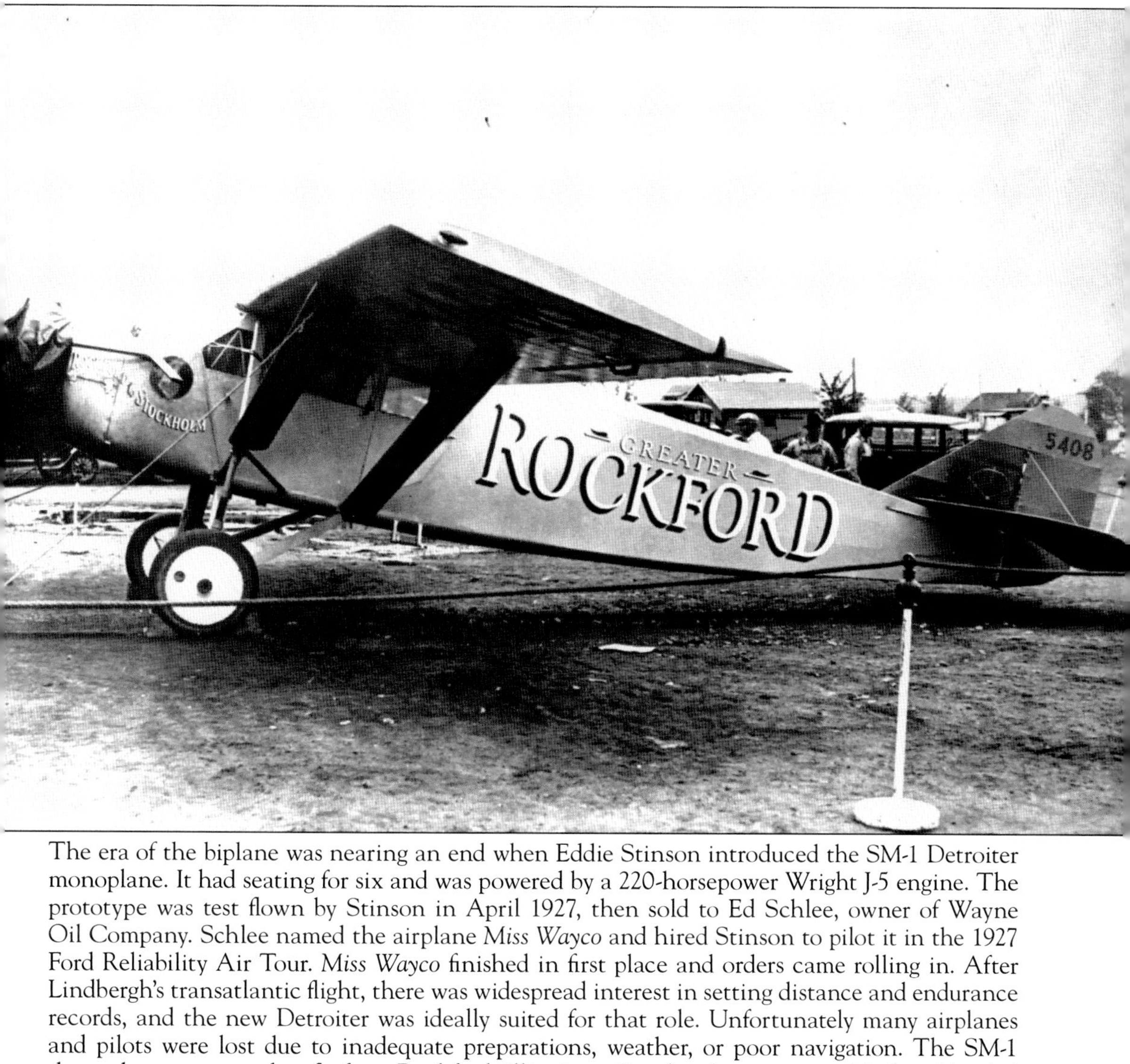

The era of the biplane was nearing an end when Eddie Stinson introduced the SM-1 Detroiter monoplane. It had seating for six and was powered by a 220-horsepower Wright J-5 engine. The prototype was test flown by Stinson in April 1927, then sold to Ed Schlee, owner of Wayne Oil Company. Schlee named the airplane *Miss Wayco* and hired Stinson to pilot it in the 1927 Ford Reliability Air Tour. *Miss Wayco* finished in first place and orders came rolling in. After Lindbergh's transatlantic flight, there was widespread interest in setting distance and endurance records, and the new Detroiter was ideally suited for that role. Unfortunately many airplanes and pilots were lost due to inadequate preparations, weather, or poor navigation. The SM-1 shown here attempted to fly from Rockford, Illinois, to Sweden, but ran out of fuel and landed on an icecap in Greenland. The record-breaking flights received lots of publicity, both good and bad, but the majority of the SM-1s delivered reliable service carrying passengers and freight. Approximately 125 Detroiters were built, including a stretched version, the SM-6, with seating for eight and powered by a 450-horsepower Pratt and Whitney Wasp.

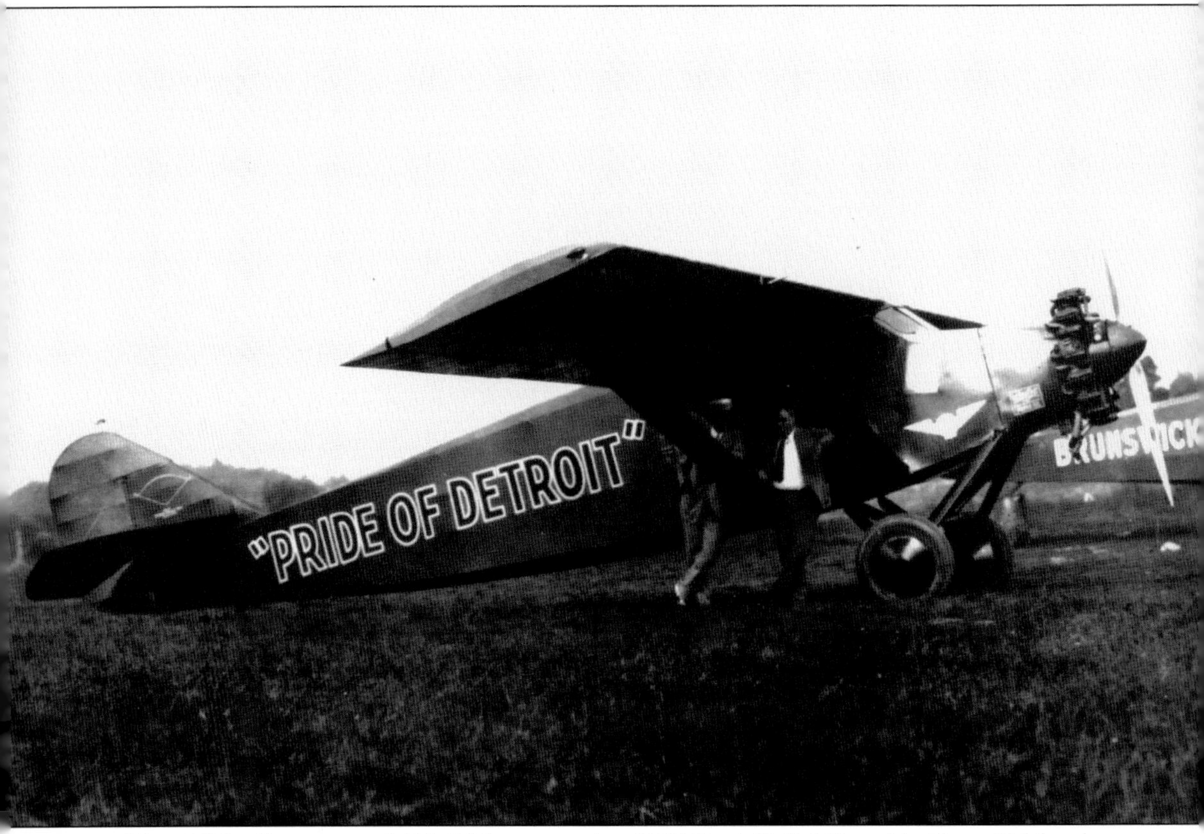

After Eddie Stinson and Ed Schlee won the 1927 Ford Reliability Air Tour, Schlee began planning a more ambitious trip with his airplane, a flight around the world. A 400-gallon fuel tank was installed in *Miss Wayco*, now named *The Pride of Detroit*, and William (Billy) Brock was named pilot. Schlee and Brock left Ford Airport on August 22, 1927, for Harbour Grace, Newfoundland, Canada. Their heavily loaded SM-1 left Harbour Grace at 5:15 a.m. on August 29 and after 23 hours and nine minutes landed safely in London. From there they flew to Munich, Baghdad, Karachi, Calcutta, Hanoi, and Shanghai, arriving in Tokyo on September 14. There they were told that the U.S. Navy had turned-down their request to station ships along their planned route across the Pacific. Tired and aware of the dangers facing them, they returned to America in a cargo ship with the Stinson safely stored on deck. Their all-yellow *The Pride of Detroit* is presently on display in the Henry Ford Museum in Dearborn.

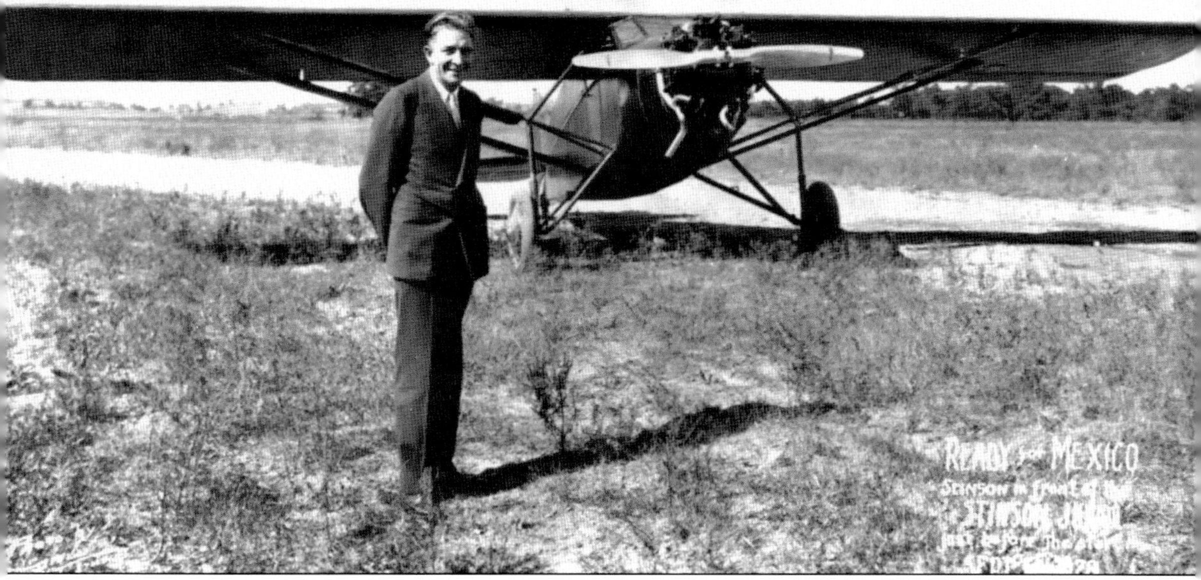

In 1928, Stinson introduced the SM-2 Junior, a smaller version of the SM-1 Detroiter. Designed by William Naylor, it was powered by a Warner Scarab engine, had seating for three or four, and sold for $8,500. The low price generated a flood of orders and eventually over 500 were built under a variety of designations. By the end of 1928, however, Stinson's Northville factory was reaching its capacity, so shares of stock were sold to raise money to build a new factory at the Detroit Wayne Industrial Airport in Wayne, Michigan. By August 1929, a 90,000-square-foot building had been built, and Stinson Aircraft moved to the new facility. Earlier that year, Stinson had been approached by Erret Cord, who wanted to buy Stinson Aircraft. Cord owned the Cord, Auburn and Dusenberg automobile companies, Lycoming Engines, and Corman Aircraft. Stinson and partner Bill Mara agreed to sell, and in September 1929, Cord bought a controlling interest in Stinson Aircraft. Two weeks later, the stock market crashed marking the start of the Great Depression. In this image, Stinson is with SM-2 Junior at Northville flying field.

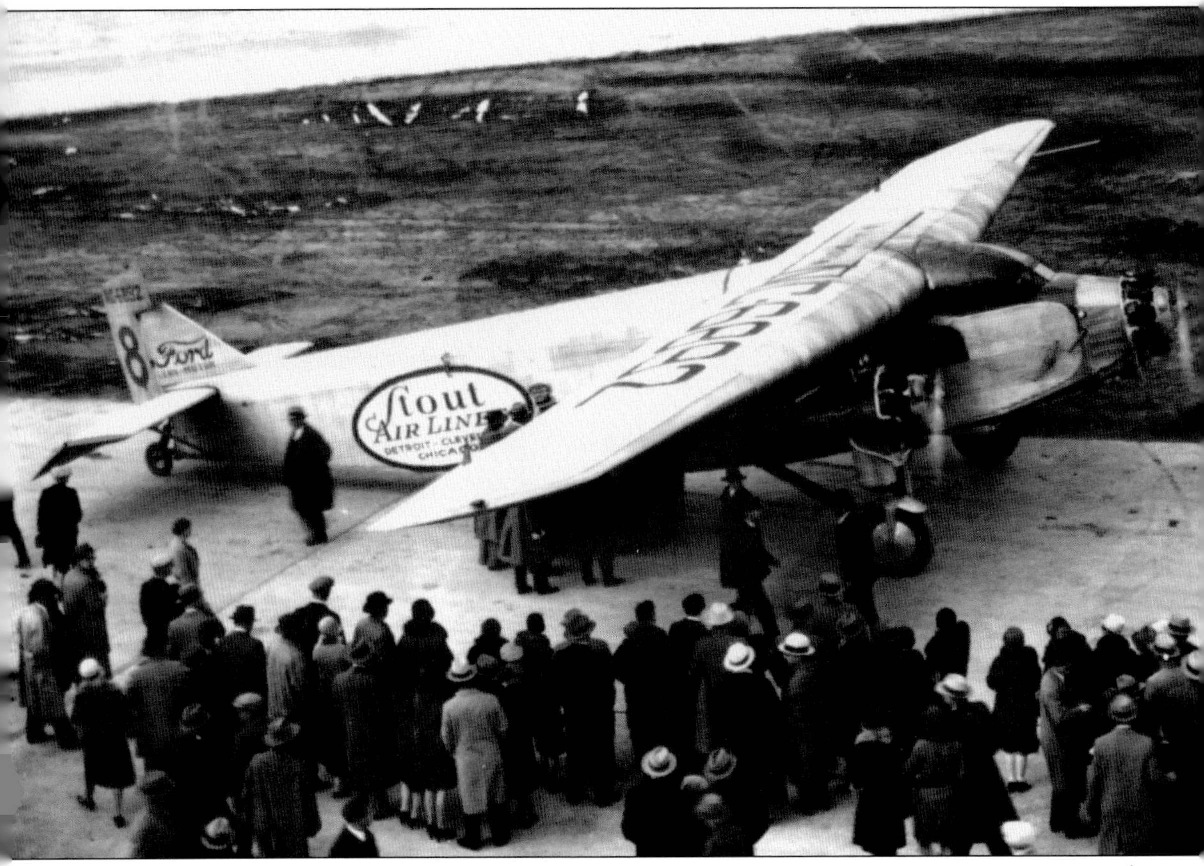

After the successful first flight of the Ford-designed 4-AT-1 Tri-Motor on June 11, 1926, Ford's Aircraft Division began preparing for production. A second 4-AT was built in October 1926, 12 more in 1927, and 38 in 1928. The last of 78 Ford 4-AT models was built in 1931. The prototype had an open cockpit, no brakes, and a tail skid, but numerous improvements were introduced during the five-year production run. A typical 12-passenger 4-AT Tri-Motor had a wingspan of 74 feet, was 50 feet long, and had an empty weight of 6,500 pounds. The power plants were 220-horsepower Wright J-5s, and the airplane sold for $42,999. Early production models were sold to Stout Air Service (SAS), the first regularly scheduled passenger airline in the United States, and the first to fly with stewards and uniformed pilots. Initially SAS flew between Dearborn and Grand Rapids, Michigan, but later expanded the route to include Cleveland and Chicago.

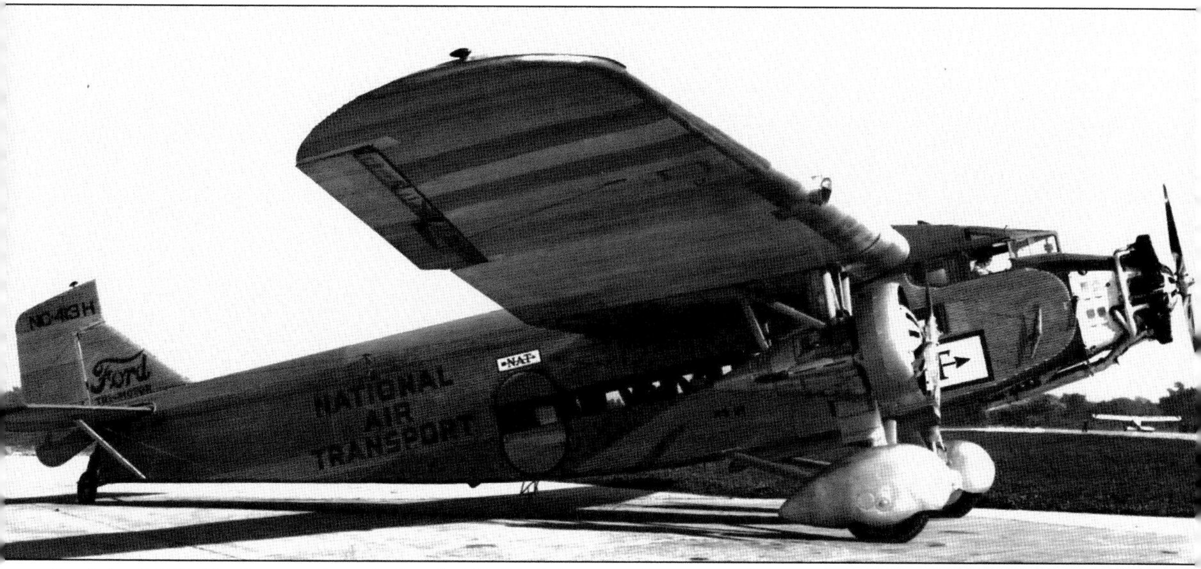

In July 1928, Ford introduced the 5-AT Tri-Motor, a larger and more powerful version of the 4-AT. The improvements had been suggested by Charles Lindbergh, technical advisor to Transcontinental Air Transport (TAT). The 5-AT carried 14 passengers, had more wing area than the 4-AT, and an empty weight of 7,600 pounds. The engines were 450-horsepower Pratt and Whitney Wasps, and the airplane sold for $55,000. Seventeen 5-ATs were built in 1928, 68 in 1929, and 2 in 1933, for a total of 119. In the summer of 1929, TAT introduced a pioneering transcontinental air-rail service. Passengers flew in a Ford Tri-Motor by day and traveled in railroad sleeping cars at night. The trip took two days but ended a year later when cross-country flights were introduced. The Ford Tri-Motors were the backbone of America's airlines in the late 1920s and early 1930s. Today a number of the classic Tri-Motors survive with six in flying condition, five preserved in aviation museums, and four presently undergoing restoration.

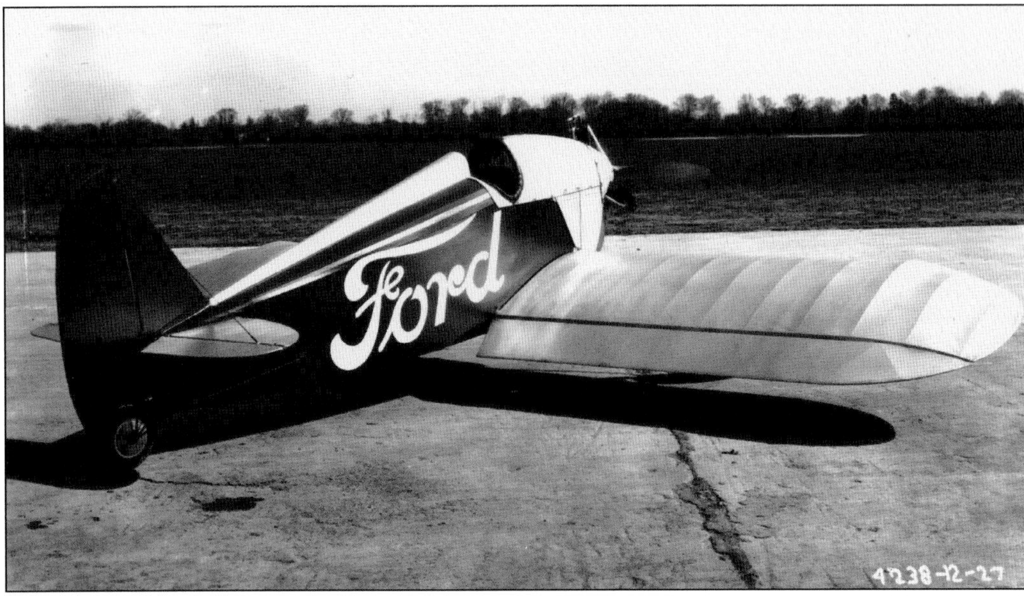

At a press conference in July 1926, Henry Ford announced the company was building a small, single-place airplane, the Ford Flivver (seen above). It was designed by Otto Koppen following suggestions from Ford test pilot Harry Brooks. The Flivver was powered by a 35-horsepower Anzani engine. Its first public appearance was at the 1926 Ford Reliability Air Tour, where it was hailed as Henry Ford's "Model T of the Air." Brooks often used the Flivver as his own personal airplane. The only other person to fly the Flivver was Charles Lindbergh during his visit to the Ford Airport on August 11, 1927 (seen below, with Lindbergh in the cockpit). That same day, Lindbergh gave Henry Ford his first airplane ride, with Ford sitting on a makeshift seat in the *The Spirit of St. Louis.*

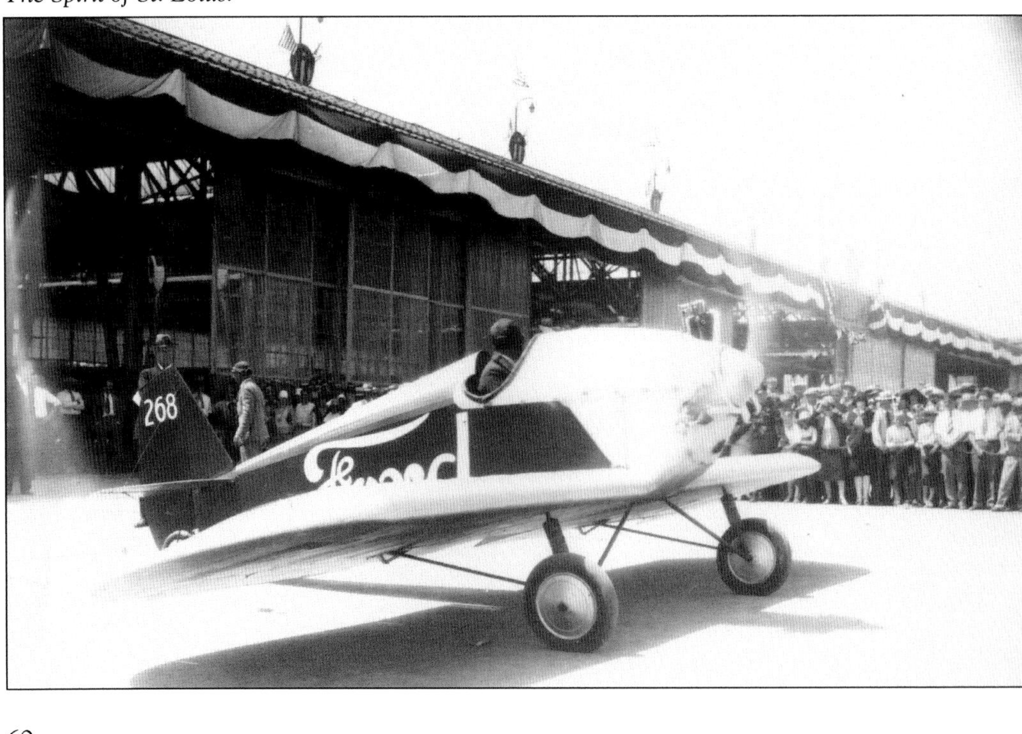

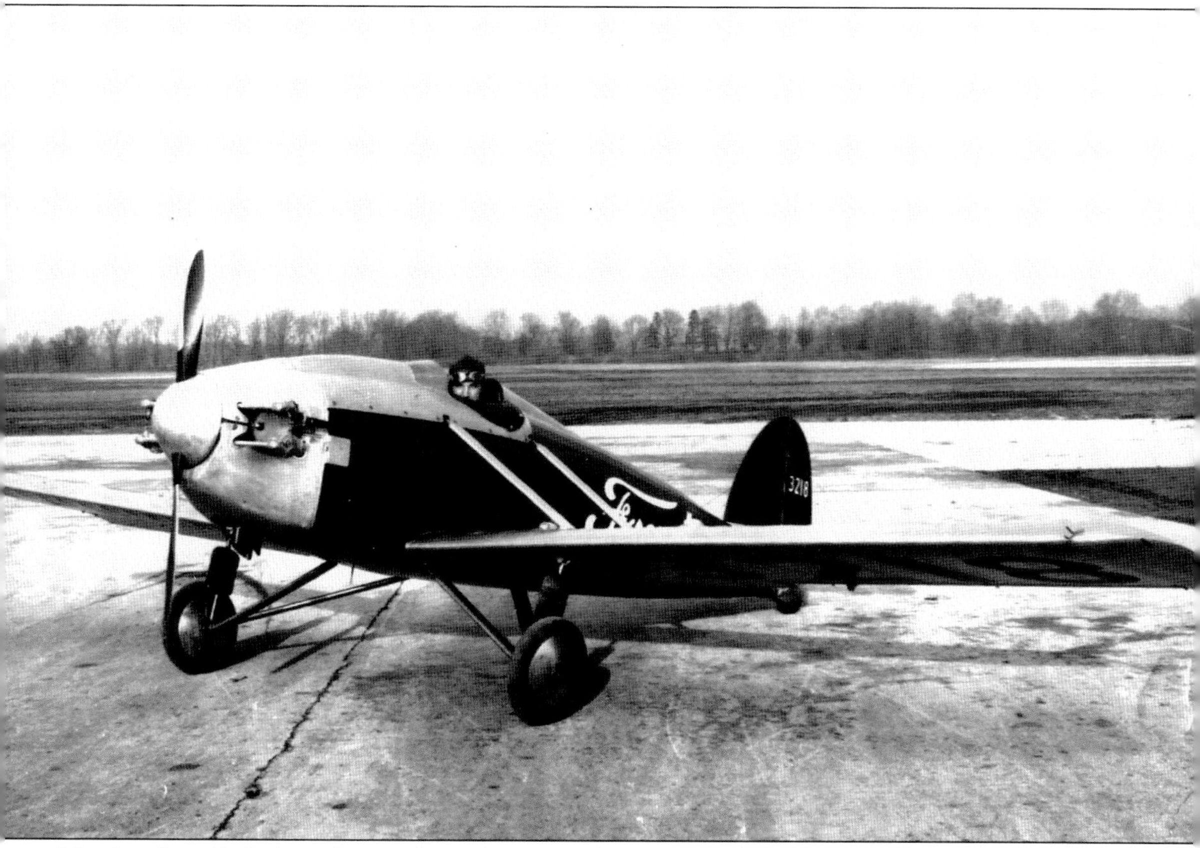

After Lindbergh's flight to Paris, other pilots wanted to set records, and Ford test pilot Brooks was no exception. He asked Koppen to design a larger Flivver capable of breaking existing records for light airplanes. The new design weighed 550 pounds and was powered by a 29-horsepower, two-cylinder, air-cooled engine (seen here, with Brooks in the cockpit) designed by Ford engineer Harold Hicks. In January 1928, Brooks took off from the Ford Airport for a nonstop flight to Miami. Adverse weather forced him to land in North Carolina, but the flight broke the United States Class "C" distance record for light airplanes. On February 21, Brooks made another attempt to reach Miami but ran low on fuel and landed on the beach near Titusville, Florida. That flight broke the world's Class "C" distance record. Four days later, shortly after taking off from the beach, spectators saw the Flivver dive into the ocean. The death of 25-year-old Brooks dampened Ford's enthusiasm for small airplanes.

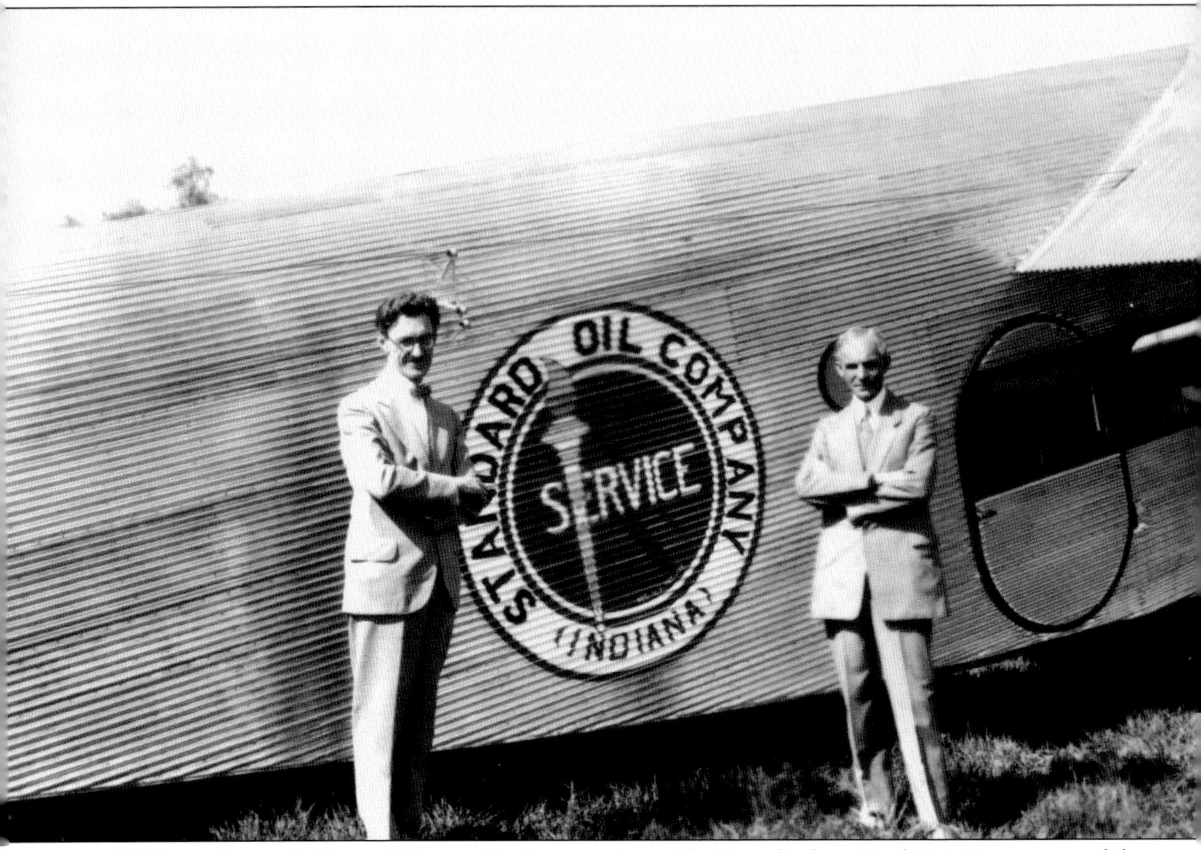

William Stout (left) and Henry Ford pose alongside *Stanolind*, a Ford Tri-Motor owned by the Standard Oil Company of Indiana. Kindred spirits, Stout and Ford were both engineers, entrepreneurs, and pioneers in their respective fields. The failure of Stout's first tri-motor design, the unsuccessful 3-AT, seems to have been forgotten when this photograph was taken. Any animosity that may have existed between the two men apparently had turned to respect following the success of the Ford-designed Tri-Motors. Stout often stated that his "greatest contribution to aviation was his successful effort to interest Henry Ford in aviation and manufacturing airplanes." If for no other reason, Stout must be recognized as one of the major pioneers in America's aviation industry.

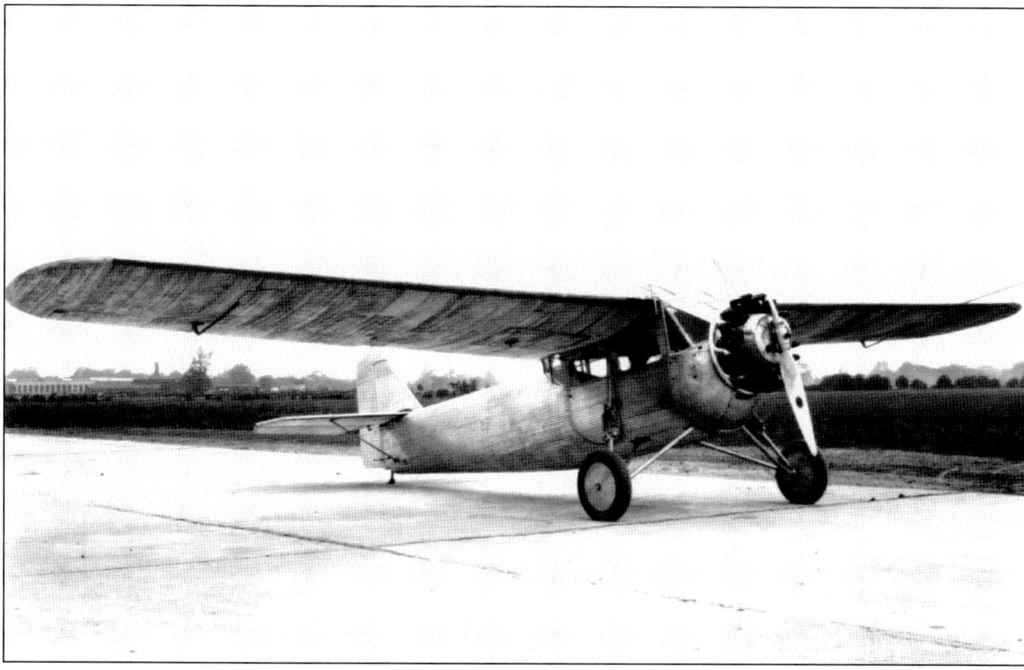

With the success of the Ford-designed Tri-Motors, Stout no longer had an active role in the program. However, Ford gave Stout free rein to design, develop, and build new airplanes, an early version of Lockheed's famous "Skunk Works." In 1927, Stout built this five-place, all-metal monoplane (seen above) designed for the executive market. Powered by a Wright J-5 engine, it was so tail-heavy that test pilot Eddie Hamilton stated that "It was too dangerous to fly." Stout's second design (seen below, with Stout at left) was a two-place, all-metal amphibian powered by two Bristol Cherub engines. Test pilot Leonard Flo was unable to coax this monstrosity into the air, and it was damaged in a high-speed taxiing accident. Both of these Stout-designed airplanes were scrapped.

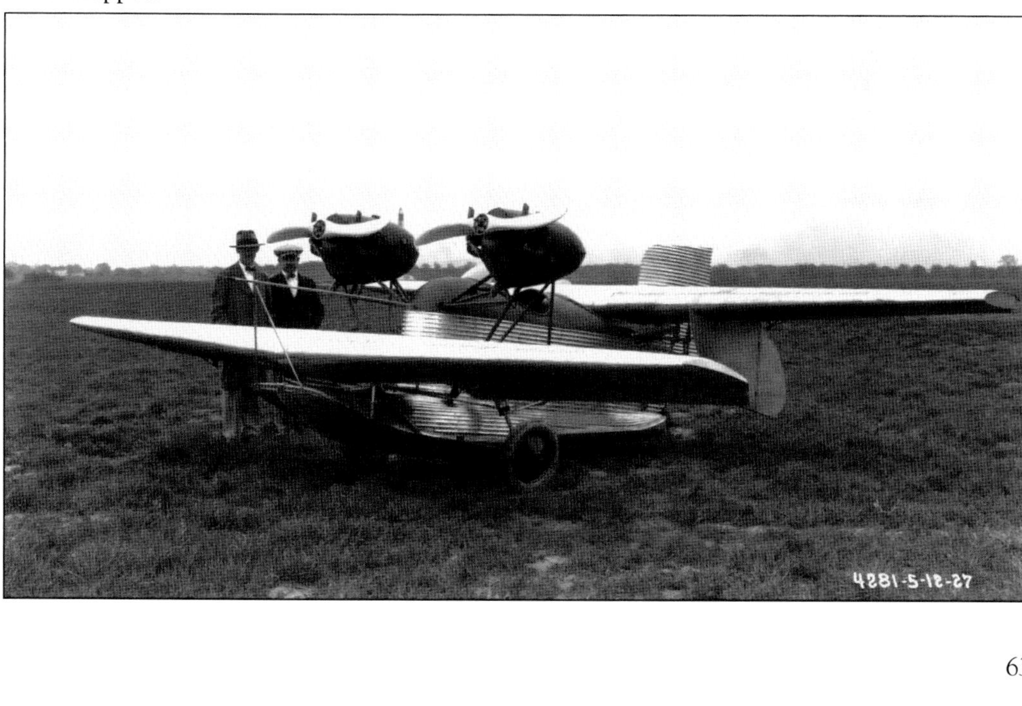

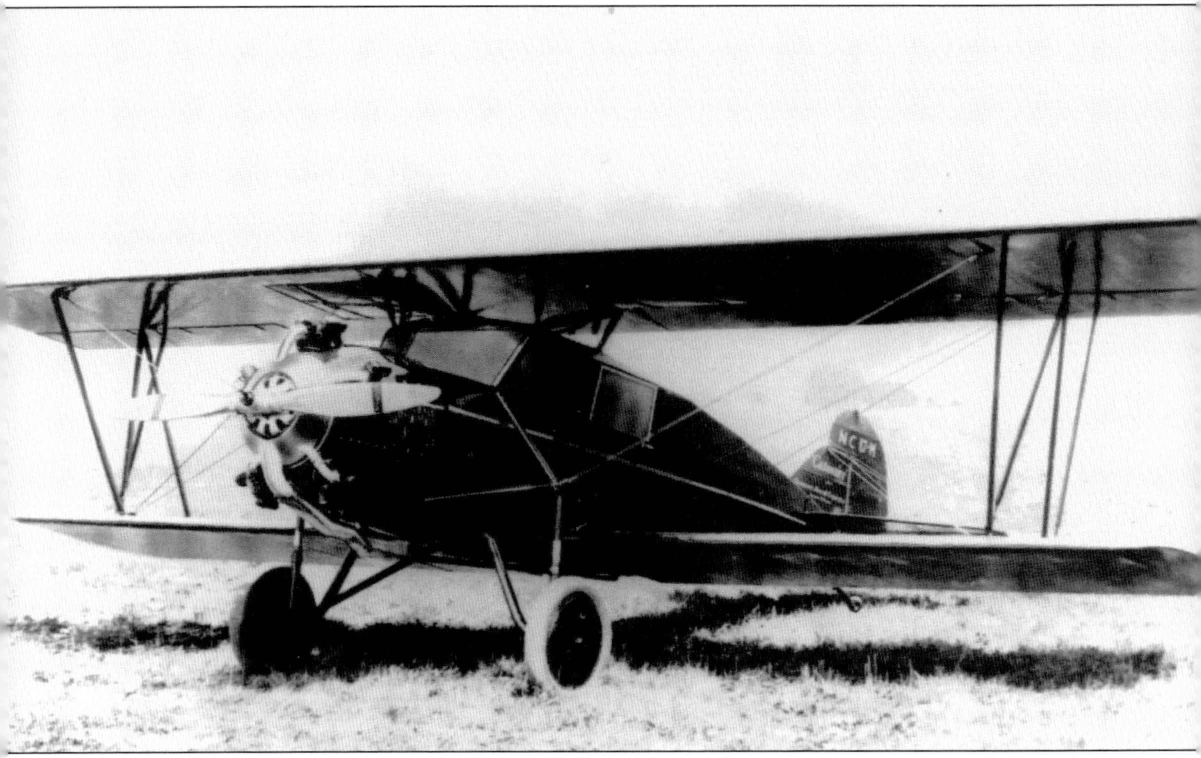

After building *Maiden Saginaw* in 1925, Walter Carr was hired by Warner Aircraft Corporation, Detroit, to flight test their new Scarab engine on his Travel Air 2000. When the tests were completed, Warner gave Carr the engine. Carr wanted to build a more successful airplane than his first attempt, so in August 1928, he and Joseph Behse, a Saginaw businessman, formed the Paramount Aircraft Corporation. For a prototype, they built a four-seat, cabin-type fuselage to which they attached the wings and tail surfaces from Carr's Travel Air. The power plant was the Scarab engine that Warner had given Carr. Eight production Cabinaires were built, powered by Warner, Continental, or Wright engines. One Cabinaire, named *The Answer*, was used by Viola Gentry to try to establish a world's endurance record. On June 28, 1929, Gentry and pilot Jack Ashcraft took off from Roosevelt Field, New York, but ran out of fuel and crashed when their refueling airplane was grounded by fog. Ashcraft was killed, and Gentry was severely injured.

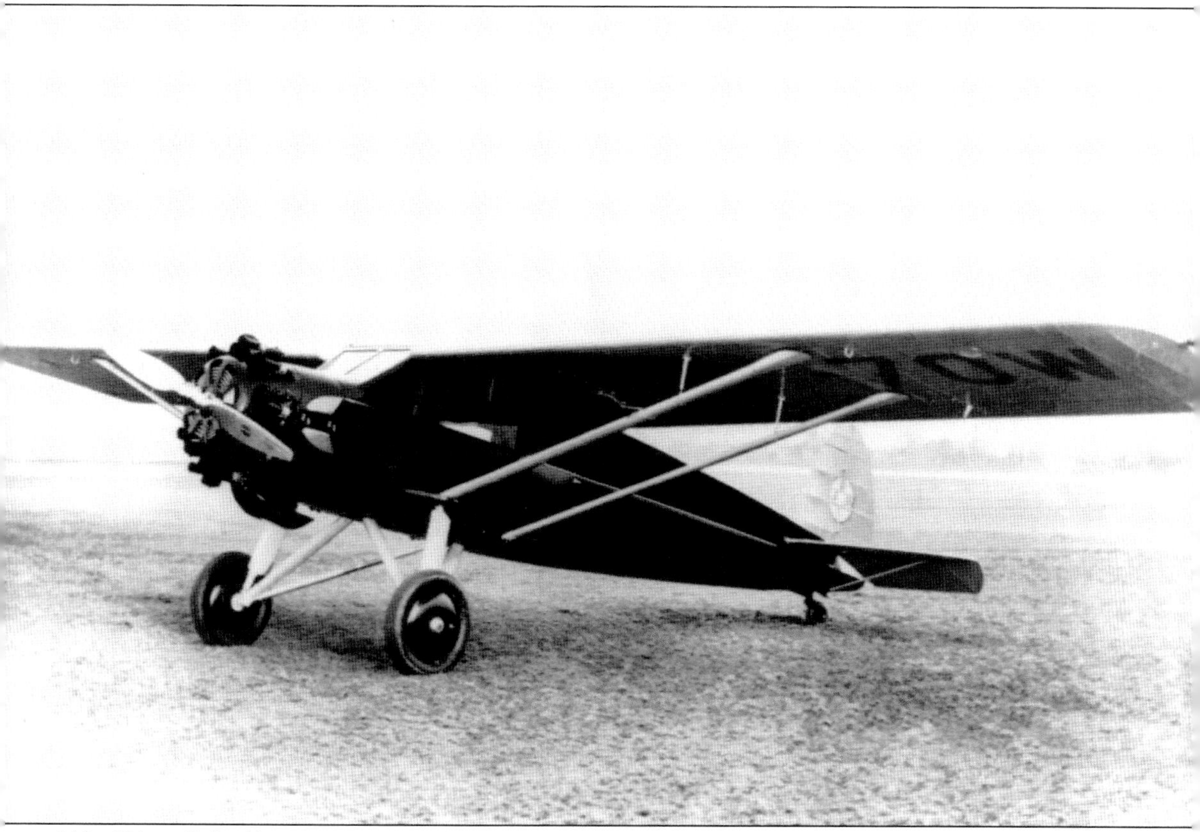

Alfred Verville had been instrumental in getting Larry Buhl established as an aircraft manufacturer. When Buhl Aircraft Company moved to Marysville, Michigan, Verville quit to form Verville Aircraft Corporation, located in the former Rickenbacker automobile factory in Detroit. Verville spent over a year designing a handsome-looking, well-appointed, four-place monoplane, the model 104 Air Coach. The prototype, powered by a 110-horsepower Warner Scarab engine, was test flown at Detroit's City Airport by William (Billy) Brock, then displayed at the 1929 Detroit Aircraft Show. The Air Coach sold for $12,000 and was available with either 165- or 225-horsepower Wright engines. But Verville had spent several years perfecting the Air Coach, and it was introduced just as the Great Depression began. Only 10 were built. One Air Coach was powered by a 225-horsepower Packard diesel engine, but on April 23, 1930, Capt. Lionell Woolson, the engine's designer, was killed flying that airplane in a weather-related accident.

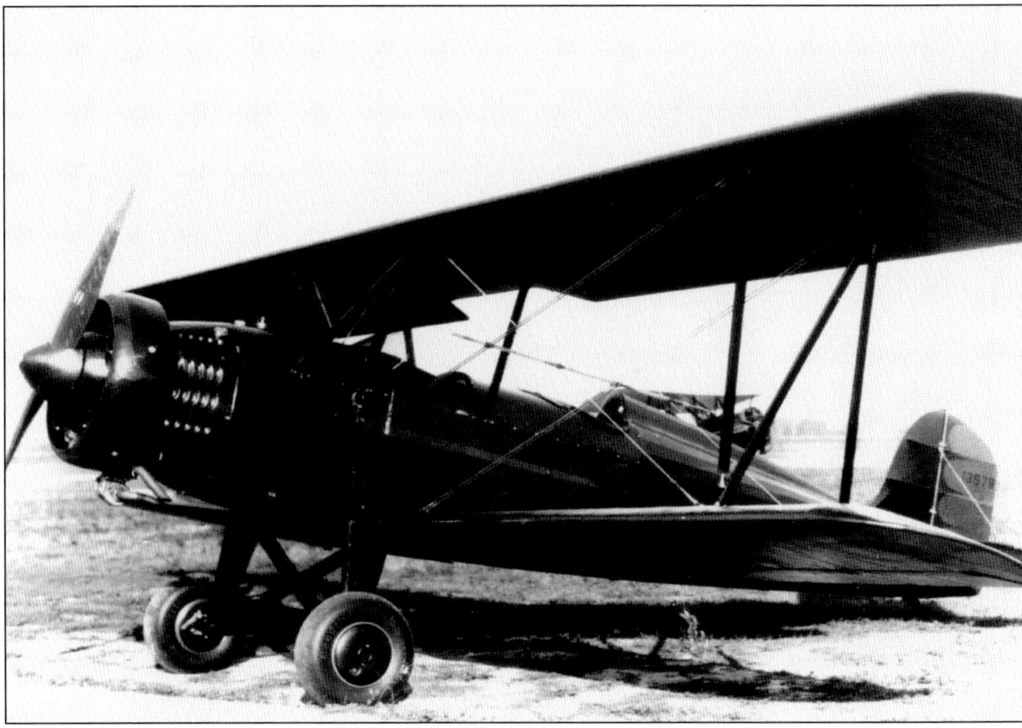

Frank Trella studied aircraft design at the University of Detroit and in 1924 built an all-wood biplane with a Ford Model T engine, but it never flew. A later Anzani-powered biplane flew well. In 1928, Trella formed Trella Aircraft and with his four brothers built this Lambert-powered Speedster. It was displayed at the 1928 Detroit Aircraft Show priced at $2,700. Only two were built.

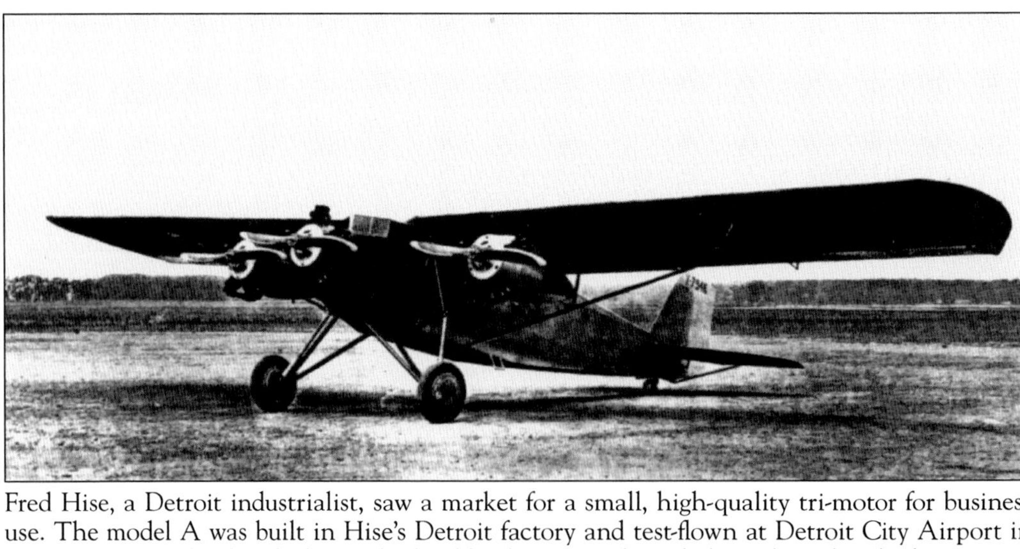

Fred Hise, a Detroit industrialist, saw a market for a small, high-quality tri-motor for business use. The model A was built in Hise's Detroit factory and test-flown at Detroit City Airport in late 1929. Painted red and silver, it had red leather seats, dome lights, silver-plated ash trays and powered by three 125-horsepower Kinner engines. Only one Hise model A was built.

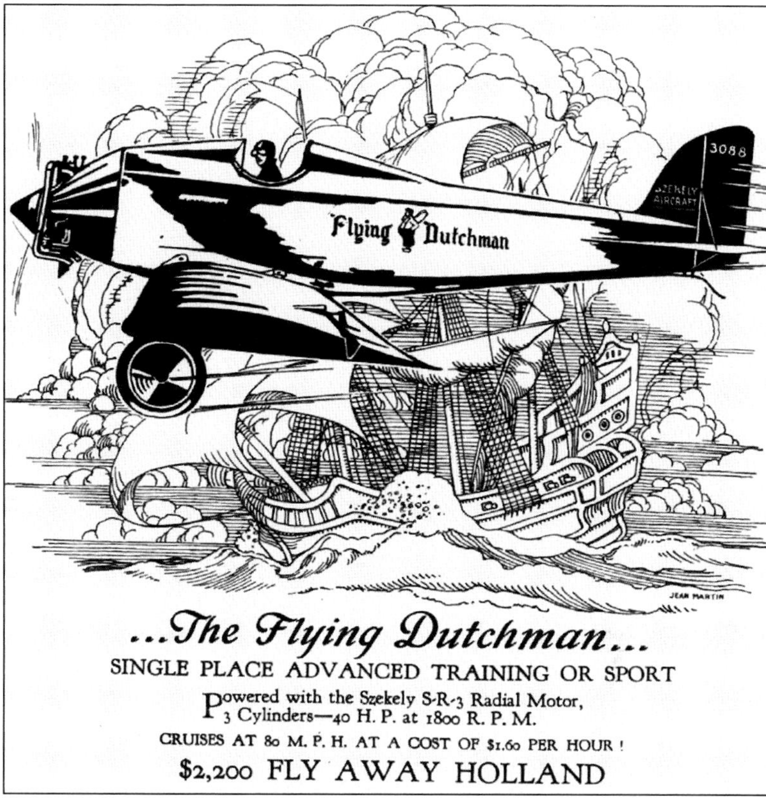

James Williams formed the Niles Aircraft Corporation, Niles, Michigan, in 1927, to build the Williams Gold Tip. The single-place monoplane, powered by an Anzani engine, was designed by Prof. Peter Altman, University of Detroit. The design rights were later acquired by Szekely Aircraft and Engine Company, Holland, Michigan. They renamed it the Flying Dutchman and powered it with a 40-horsepower Szekely engine. Approximately 20 were built in 1928 and 1929.

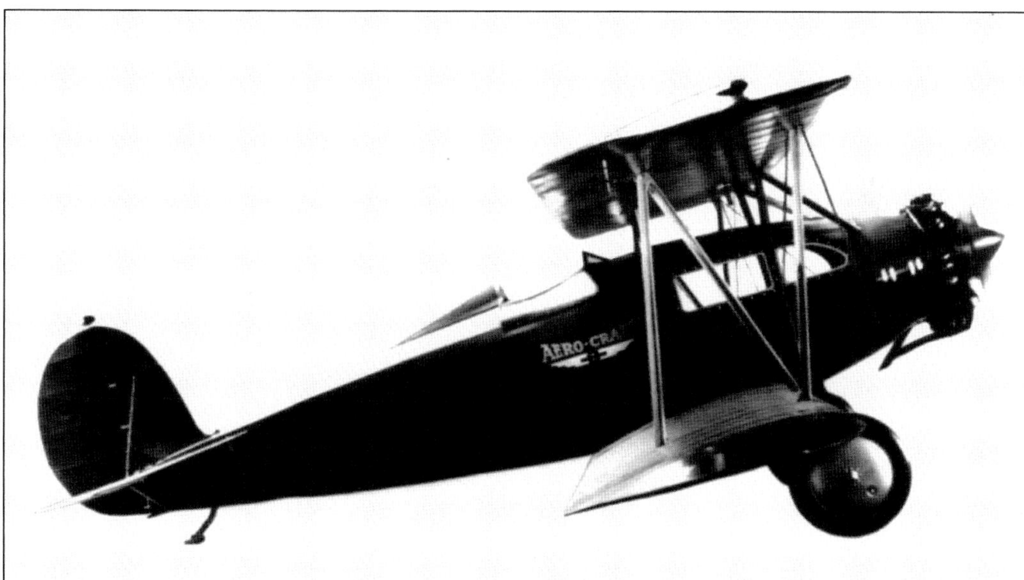

Stinson's cabin biplane inspired others to build similar designs. In 1927, the Gillis Aircraft Company, Battle Creek, Michigan, built the Gillis Crusader, a four-place cabin biplane with a Siemens engine (not shown). In 1928, Aero-Craft Manufacturing Company of Detroit built the Aero-Coupe with a Warner Scarab engine (seen here). Designed by Capt. Clement Brown, the passengers sat in he cabin but the pilot still sat in an open cockpit.

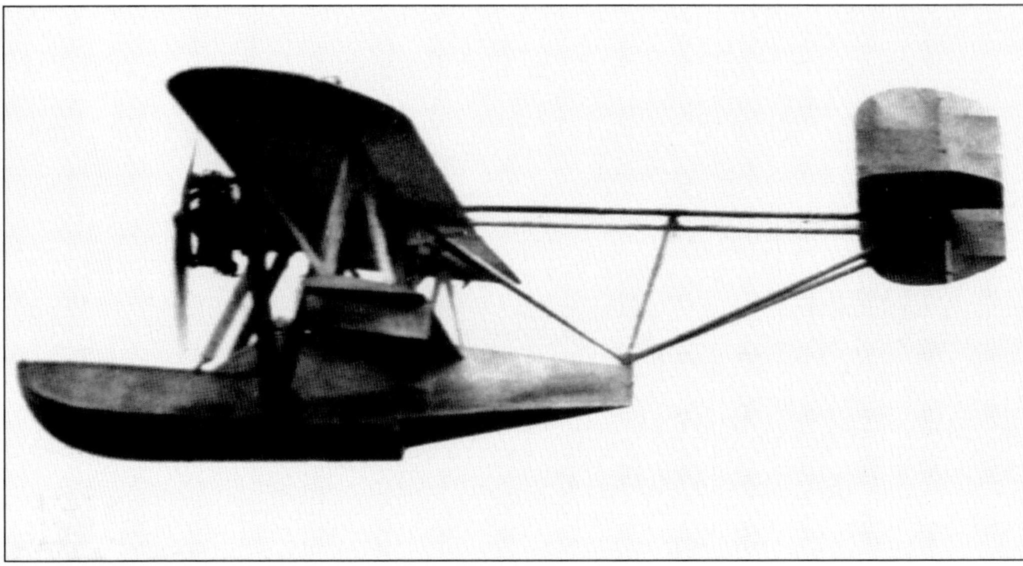

Tom Towle was born in Dayton in 1897. A Yale engineering graduate, he worked for Orville Wright, Glenn Martin, Bill Stout and, after Ford bought Stout's company, for Ford. At Ford, Towle helped design the first Ford 4-AT-1 Tri-Motor. He left Ford in 1928 to form the Towle Marine Aircraft Company, Detroit, and designed and built this two-place Beasley-Eastman flying boat (seen above) for Jim Eastman, a Detroit industrialist. Towle later designed a larger, four-place version powered by a 180-horsepower Curtiss Challenger engine. Jim Eastman then formed the Eastman Aircraft Corporation in Detroit to manufacture the new design. Some 20 E-2 Sea Rovers were built (seen below), including amphibious versions known as Sea Pirates. Eastman sold the company to the Detroit Aircraft Corporation in July 1929.

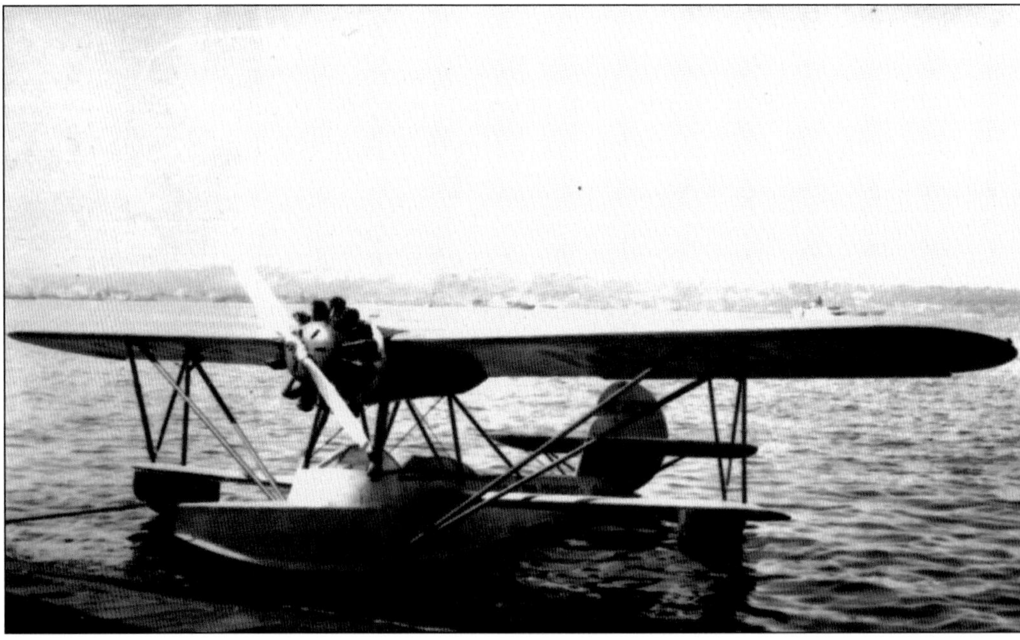

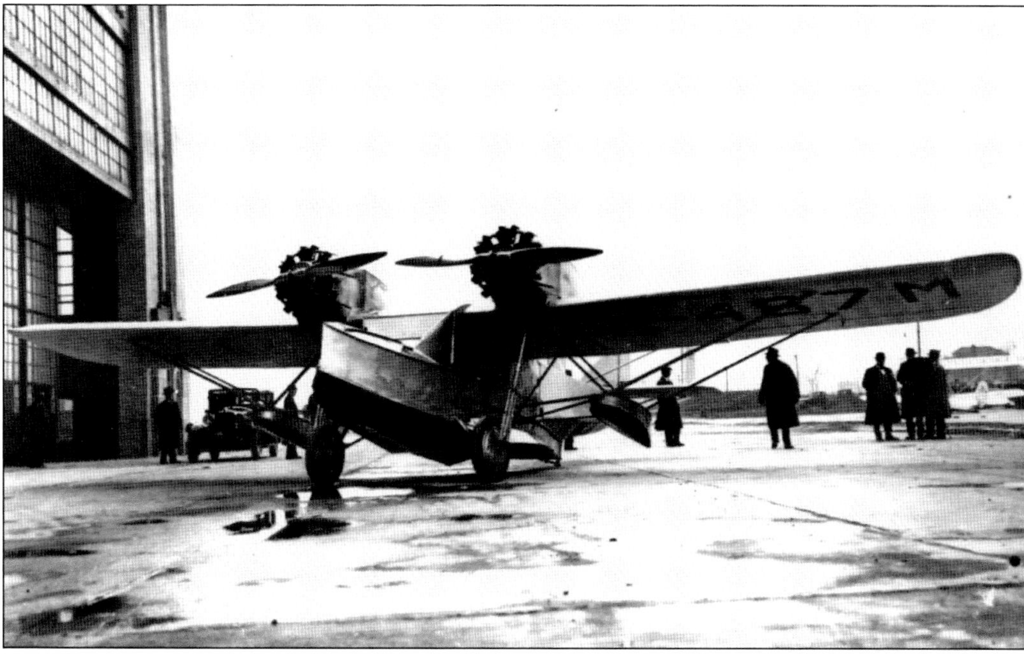

In 1928, Henry McCarroll formed the Cadillac Aircraft Corporation and built a four place, all-wood amphibian (not shown) in the ex-Stinson factory in Northville, Michigan. Flown in late 1929, it was known as the Cadillac *Voyageur* or McCarroll *Duoplane* and had two Kinner engines. In 1930, Thomas Lamphier formed Issoudun Aircraft Manufacturing Corporation and built a four place, all-metal amphibian, the Issoudun H-23 (seen here) in the same factory.

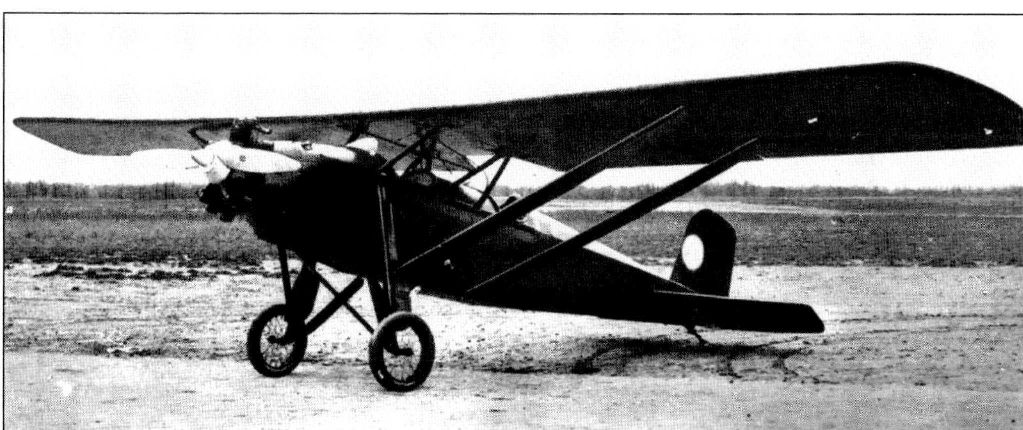

Price $2,875. Flyaway Our Field, Lansing, Mich.
Two Place-Tandem Cockpit (after the Army and Navy designed planes which are the result of millions of dollars experimental work).

The "Lindbergh Boom" created a demand for training aircraft, so Prof. Peter Altman of the University of Detroit designed a two-place monoplane, the Derickson Air Trainer. In 1928, the design rights were sold to brothers Donald and John Thomas, who formed Capital Aircraft Company in Lansing. Capital Aircraft built one Air Trainer powered by a 60-horsepower LeBlond engine (seen here), then sold the design to Royal Aircraft, Royal Oak, Michigan.

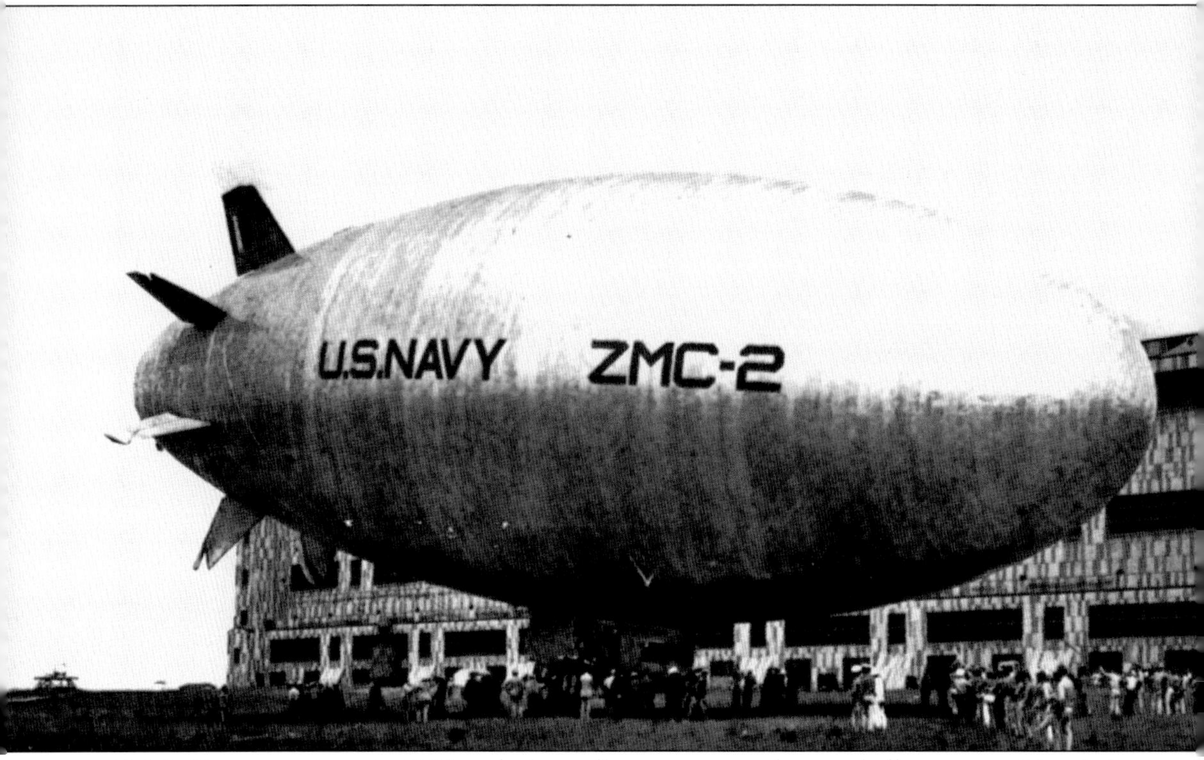

Ralph Upson was born in 1888 and after college won several major balloon races. That led to a job with the Goodyear Rubber Company teaching military cadets to fly balloons. After he became chief engineer of Goodyear's Airship Division, Upson's proposal to build an all-metal airship was turned down by management. He resigned and moved to Detroit where he obtained financial backing and formed Aircraft Development Corporation (ADC). After several years developing the concept, the U.S. Navy awarded ADC a $300,000 contract to build a metal-clad airship. A hangar was erected on Grosse Ile, Michigan, in which to build the airship, which was 50 feet in diameter and 150 feet long. A unique riveting machine was designed to join the .009 inch thick Alclad skin that formed the hull, which was built in two halves, front and rear, then joined together. First flight was on August 19, 1929. Designated ZMC-2 by the U.S. Navy, the airship was delivered to Lakehurst Naval Air Station, New Jersey, and for 10 years was used to train airship pilots.

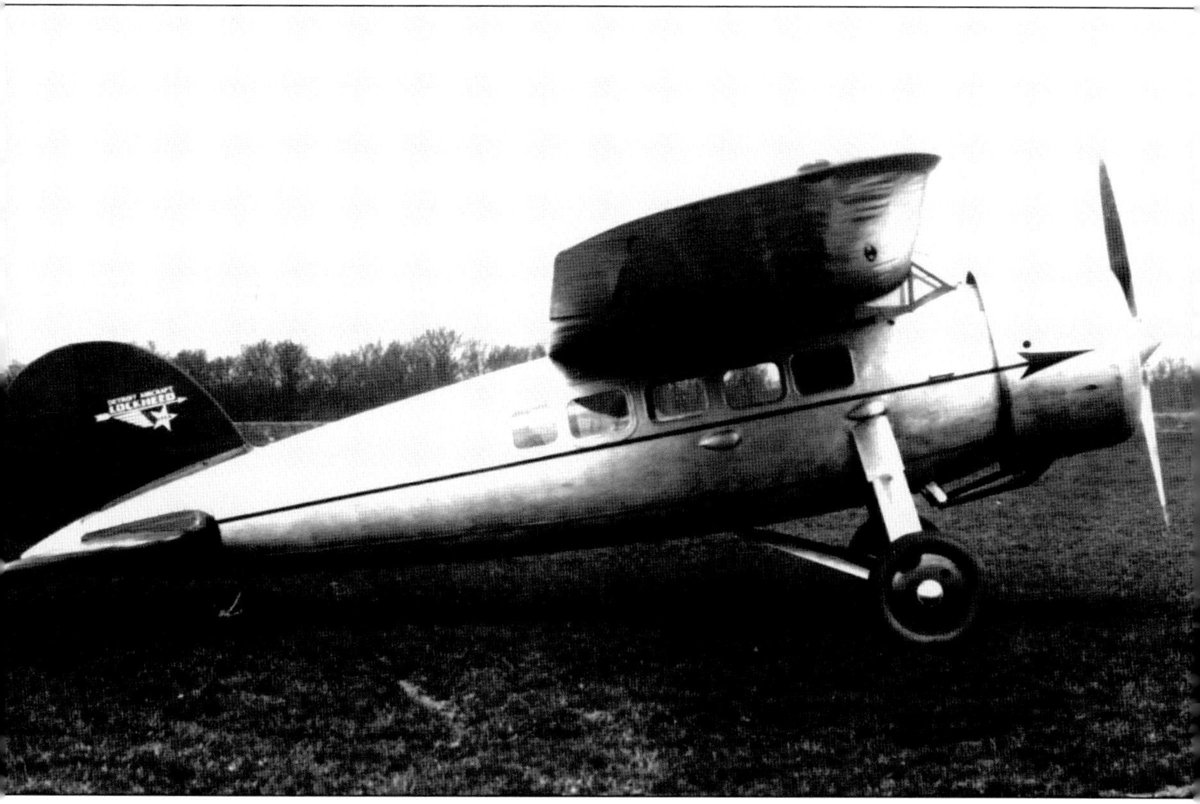

In early 1929, ADC was reorganized and became the Detroit Aircraft Corporation (DAC). The original company that built the ZMC-2 airship (ADC) was now a division of DAC. The new conglomerate planned to expand into aircraft manufacturing and its goal was to become a "General Motors of the Air," and by June 1929 had acquired controlling interest in several well-established aircraft companies including Lockheed; Ryan; Eastman; Gliders Incorporated; Grosse Ile airport, near Detroit; Parks Air College; and other entities. The crown jewel of DAC's acquisitions was Lockheed Aircraft, whose six-passenger Vega was selling briskly. However, the Vega had a molded plywood fuselage that was time-consuming to build, so DAC engineers designed a metal fuselage as a replacement. Nine metal fuselage Vegas were assembled in Detroit in 1930 with wooden wings and tail surfaces built by Lockheed in California. All Detroit-built Vegas were test flown at DAC's Grosse Ile airport.

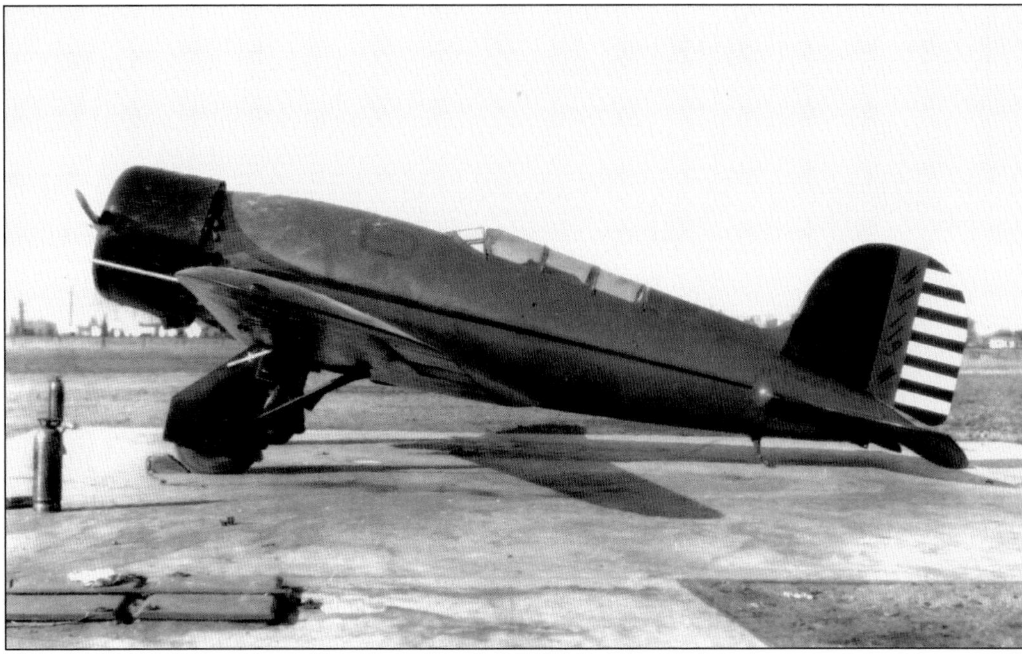

In addition to the nine high-wing Vegas, Detroit-Lockheed also built two low-wing Sirius monoplanes with metal fuselages. They were mated to California-built wooden wings with fixed landing gears. Both airplanes were later fitted with Lockheed Altair wings that had retractable landing gears. One was sold to the U.S. Army Air Corps as the Y1C-23 (seen above) and used for VIP transportation. The other was sold to the U.S. Navy as the XRO-1 (seen below). Despite the depressed economy, Lockheed airplanes were selling well but other divisions of the Detroit Aircraft Corporation were losing money, so on October 27, 1931, DAC halted all aircraft production and declared bankruptcy. In June 1932, the Lockheed division was bought for $40,000 by a group headed by Robert Gross, a San Francisco investment broker. That small investment has grown into today's aerospace giant, Lockheed Martin.

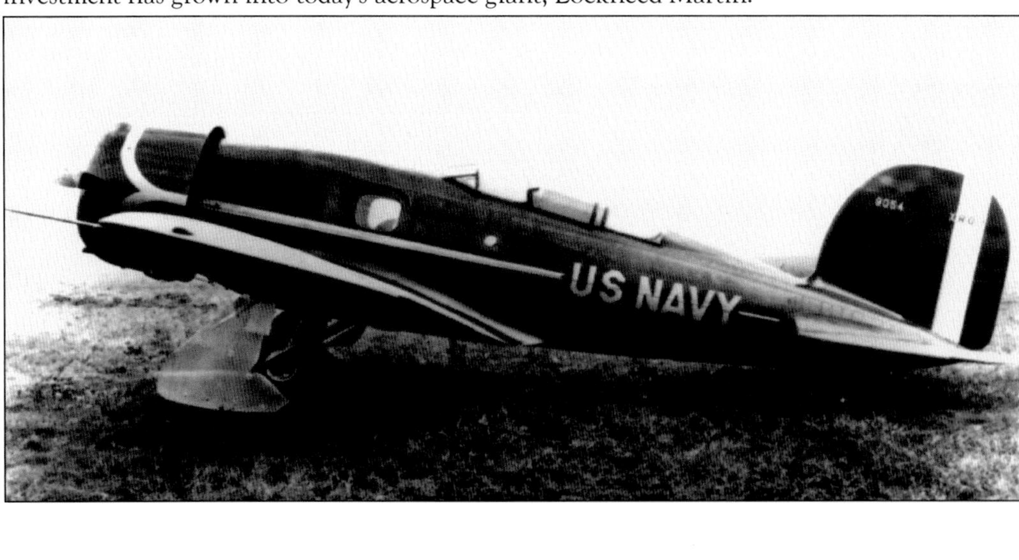

Five

The Depression Years 1930–1938

The Lindbergh Boom came to an end with the collapse of the stock market in October 1929, and as the economy worsened, there was little demand for new airplanes. The huge Detroit Aircraft Corporation folded in October 1931, and Buhl Aircraft Company closed in February 1933. Ford's Aircraft Division lasted until April 1936. Stinson Aircraft was one of the few survivors and continued selling airplanes despite the depressed economy.

Before the full impact of the Great Depression was felt, southeastern Michigan continued to be a center of aviation activity. The All-American Aircraft Show, held in Detroit in April 1930, attracted 200,000 visitors with 80 airplanes on display. The last of those annual shows was held in 1933. In July 1930, the Cirrus Engine Company of Marysville sponsored the All-American Flying Derby, a 5,541-mile race around the United States. The last of the annual Ford Reliability Air Tours was held in 1931. But despite the weak economy, there were many individuals who wanted to fly, and they found the answer by flying gliders. As a result, glider clubs and glider manufacturers became active during the Great Depression years.

The previous decade saw many technological and aerodynamic advances introduced but they produced only minor gains when applied to existing airplanes. However, when applied collectively to new designs they produced airplanes that were a quantum leap beyond existing airplanes. Boeing's 247 and the Douglas DC-2, both introduced in the early 1930s, marked the beginning of a new era for aircraft designers. Furthermore, to stimulate progress in the industry, Eugene Vidal, director of the Bureau of Air Commerce, announced in 1933 a competition to design and build a $700 private airplane. It was an unrealistic goal, but it did produce a number of innovative designs.

There were 12 million unemployed in America during the peak years of the Great Depression, and the aircraft industry was one of the last to recover. But by 1938, war clouds were gathering in Europe creating a demand for new military aircraft, both in this country and overseas, and the American aircraft industry began a slow recovery.

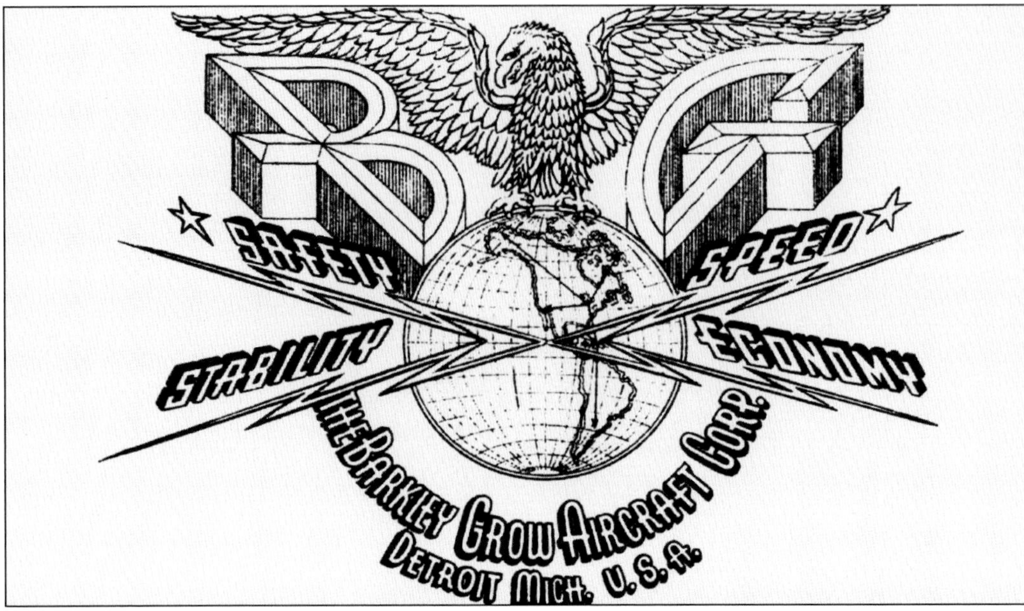

The Barkley-Grow Aircraft Corporation, Detroit, used this elaborate logo on the vertical fin of its all-metal, twin-engine transports. The logo included a globe showing the western hemisphere, an American eagle, lightning bolts, a *B* and G in large block letters, and the words safety, stability, speed, and economy radiating from the globe. This illustration came from the Barkley-Grow sales brochure.

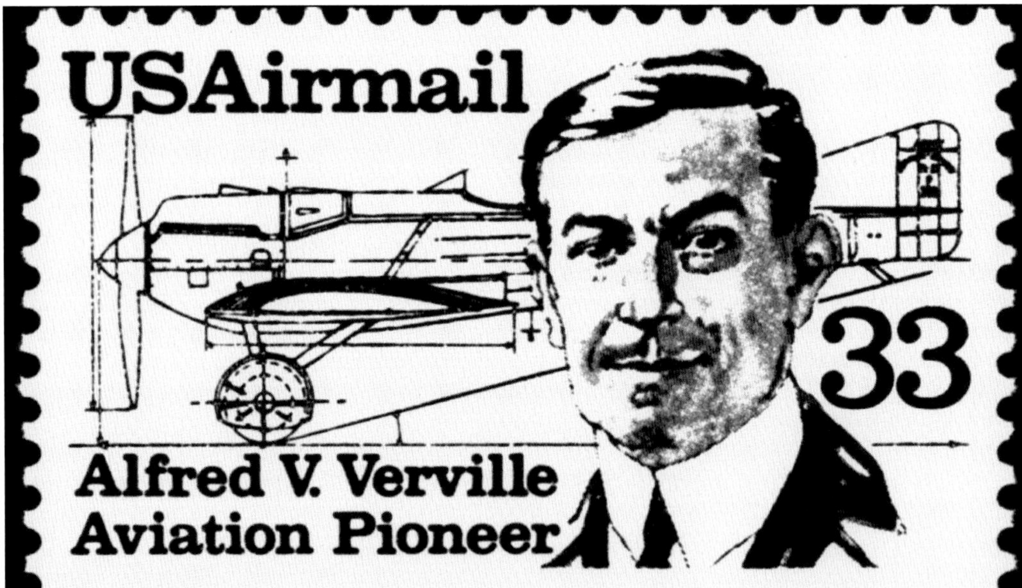

Michigan-born Alfred Verville is one of the few engineer/designers to be honored on a United States postage stamp. This 33¢ airmail was issued on February 13, 1985. Behind Verville's portrait appears a drawing of his 1922 R-3 racer. With cantilevered wings and retractable landing gear, it was far advanced over contemporary aircraft and won the 1924 Pulitzer Trophy Race with a speed of 216.55 miles per hour.

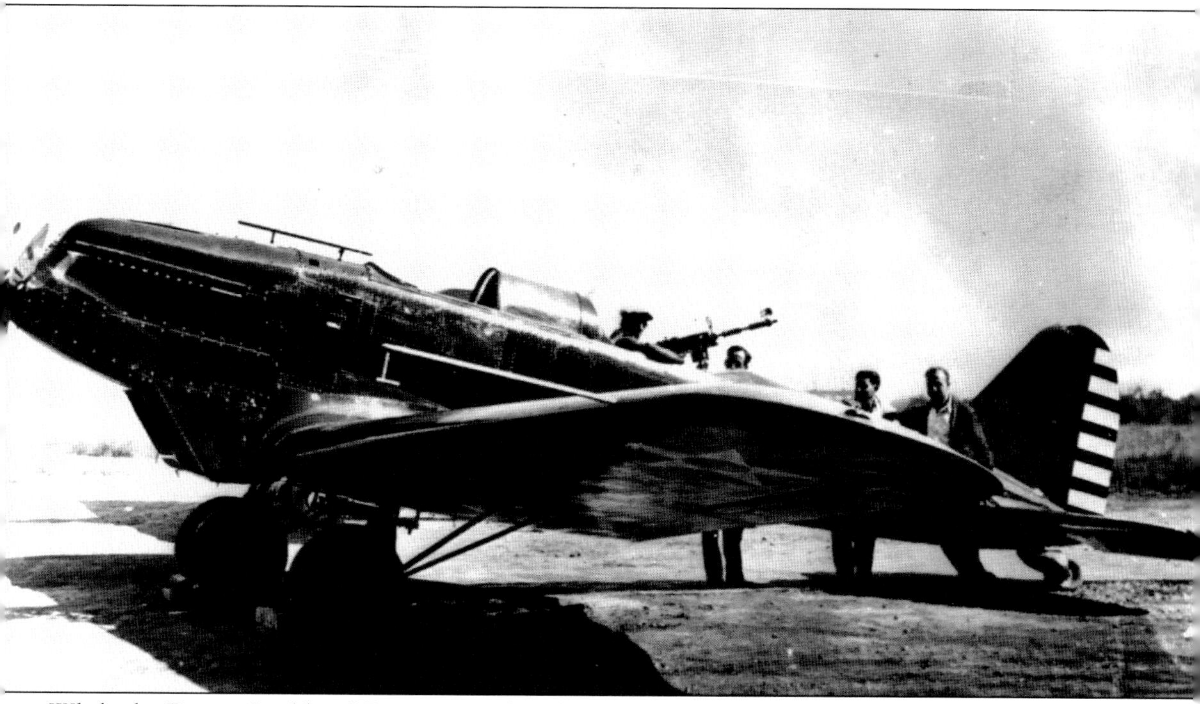

While the Detroit-Lockheed Division was building Lockheed Vegas for the commercial market, engineer Robert Woods was designing this two-place fighter plane for the U.S. Army Air Corps. It featured an all-metal fuselage with a 600-horsepower Curtiss Conqueror engine and used a California-built Lockheed Altair wing. First flight was made by Vance Breese in the summer of 1931 (seen here at far right). The temporary XP-900 designation was changed to YP-24 after delivery to the U.S. Army Air Corps in September 1931. But when a U.S. Army Air Corps test pilot had difficulties lowering the landing gear, he bailed out, and the YP-24 was destroyed. Because its performance was superior to most contemporary military aircraft, the U.S. Army Air Corps ordered 10 more but Detroit Aircraft had declared bankruptcy, so the contract was awarded to Consolidated Aircraft. Consolidated eventually built over 50 airplanes based on Detroit-Lockheed's original design. Engineer Woods went to work for Bell Aircraft where he designed the P-39 Airacobra and led the team that designed the Bell X-1, the world's first supersonic airplane.

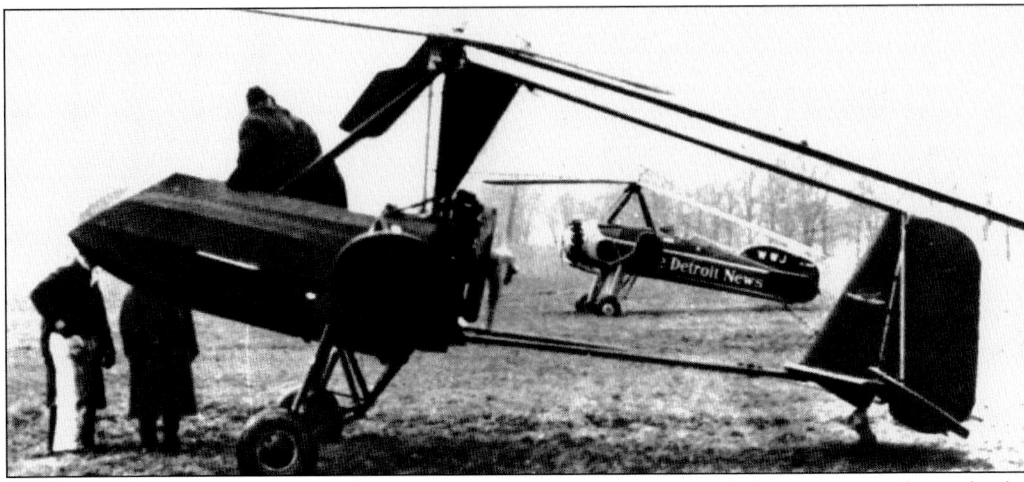

The autogiro was invented in 1923 by a Spanish engineer, Juan de la Cierva, and was hailed as an aviation breakthrough. Unlike the helicopter, a later development, the autogiro's rotor was not power-driven but was a free rotor that provided a generous amount of lift resulting in very short takeoffs and landings. Buhl Aircraft Company acquired a license to build an autogiro, and Etienne Dormoy designed this two-seat pusher first flown in December 1931 (seen above, with the Detroit News' Pitcairn autogiro in the background). Dormoy also designed two sporty-looking monoplanes during this period. The two-place CA-1WA Airster (seen below) was powered by a 420-horsepower Pratt and Whitney Wasp. A very similar-looking, single-seat CA-1 Airster was designed to carry mail and had a 300-horsepower Pratt and Whitney Wasp engine.

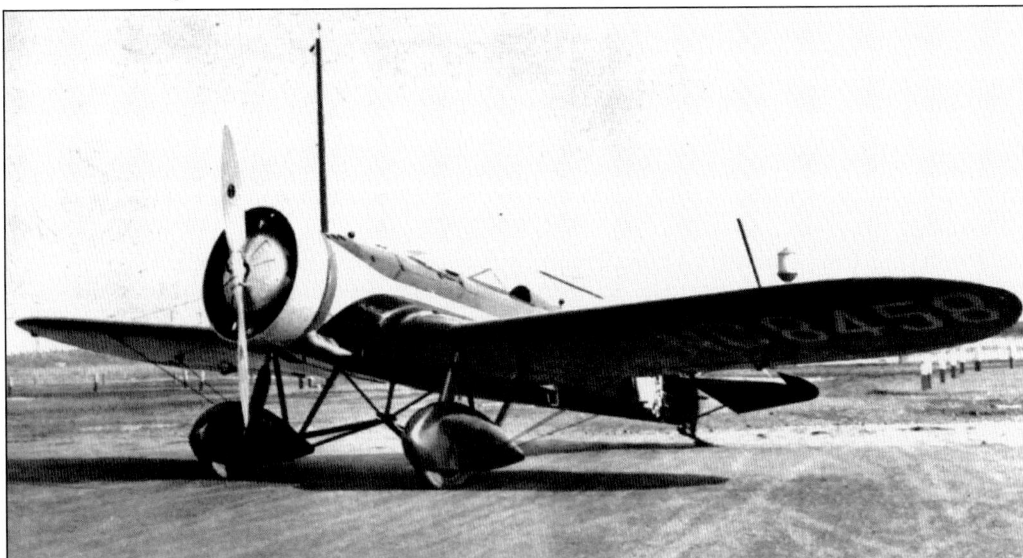

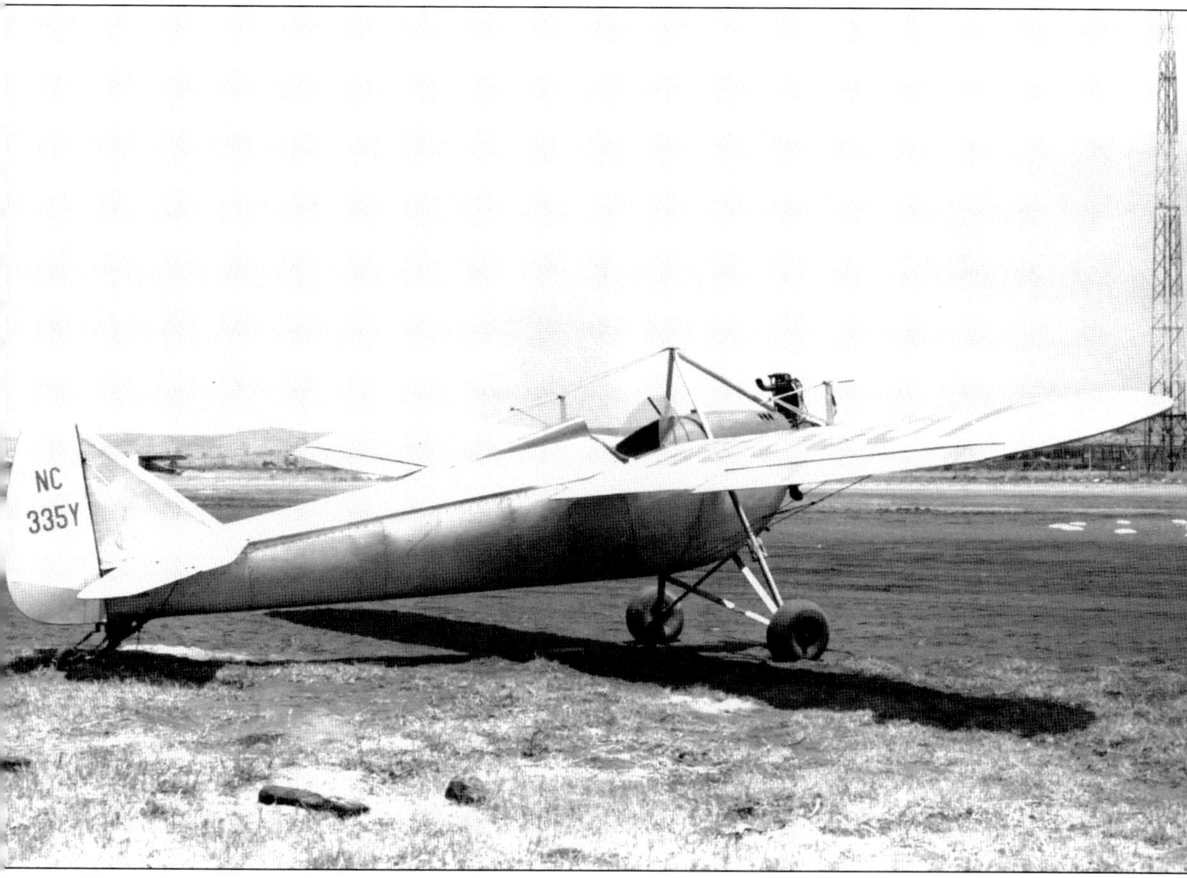

With the worsening economy, there was little demand for the big Buhl Airsedans, but president Larry Buhl saw one last chance to remain in business. He asked engineer Dormoy to design a single-seat monoplane that would appeal to pilots who wanted to fly at minimum cost. The original design had a steel-tube fuselage and a 26-horsepower Aeronca engine, but Dormoy later designed a new fuselage with two aluminum half-shells, left and right, that could be formed in the Buhl Stamping Plant. He also switched to a 45-horsepower Szekely engine, built in Holland, Michigan. The LA-1 Bull Pup was introduced in March 1931 at the Detroit Aircraft Show, priced at $1,250. Buhl received 290 orders, so a new factory west of Marysville was constructed in which to build the Bull Pup. Production began that summer, but many of the original orders were canceled, and only 100 were built. Buhl Aircraft Company closed down in February 1933, and Dormoy left to work for Boeing and later Consolidated Aircraft. He died in 1959.

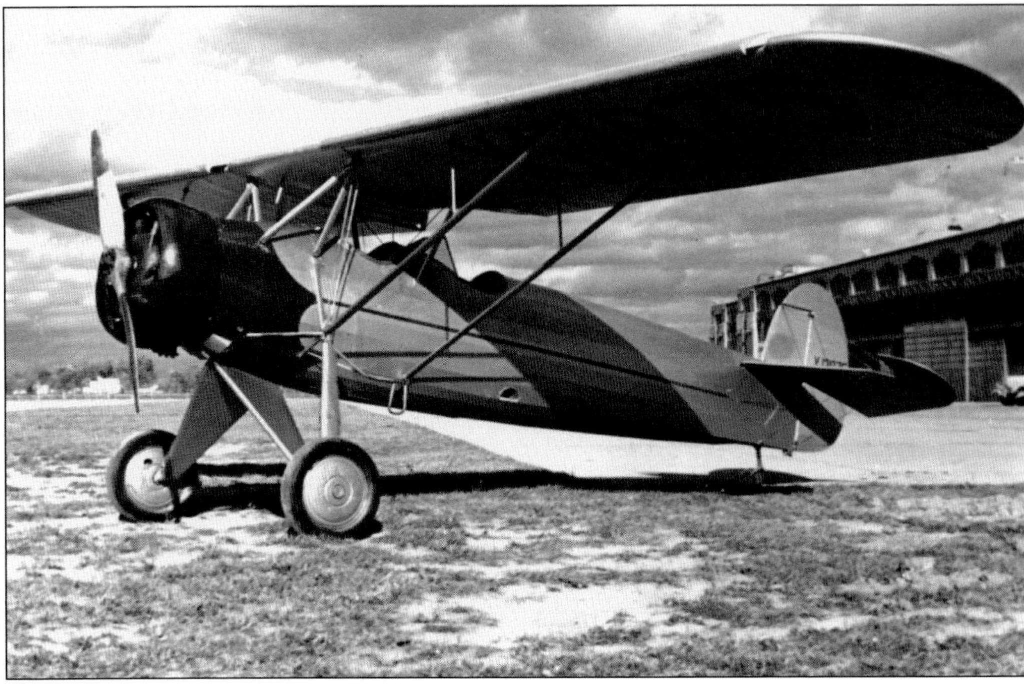

Engineer Clifford Jackson came to Marysville in 1927 to work for Buhl Aircraft Company. After Buhl Aircraft closed, he built this two-place trainer, the Jackson O-2, first flown in January 1933. Jackson went to work for Stinson Aircraft and in 1941 sold the O-2 to Meyers Aircraft in Tecumseh, Michigan. Meyers planned to put the O-2 into production but never did. Cliff Jackson died in 1992.

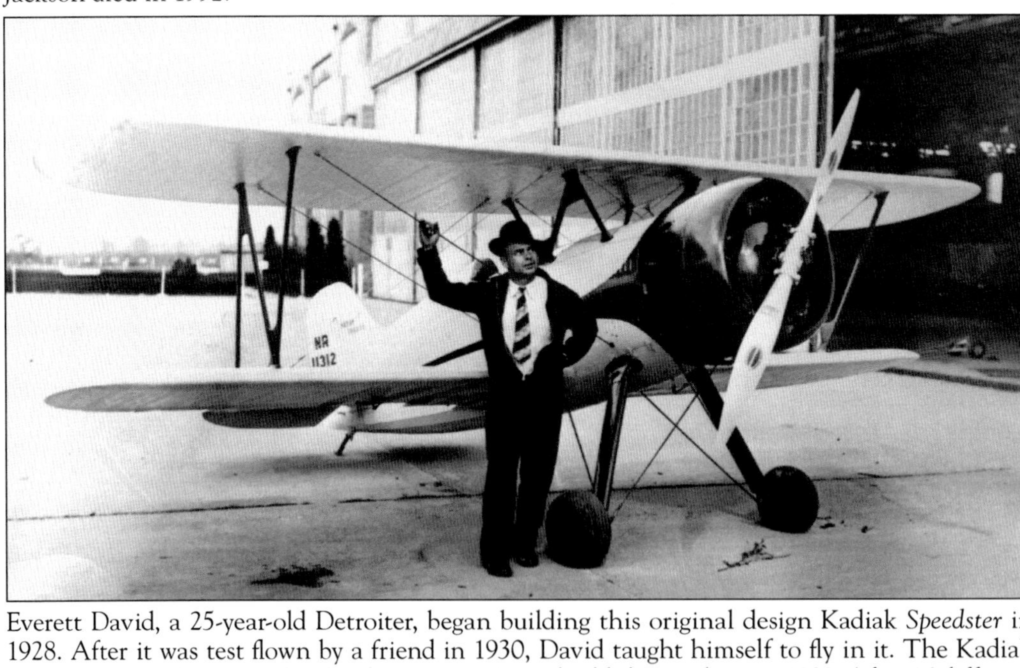

Everett David, a 25-year-old Detroiter, began building this original design Kadiak *Speedster* in 1928. After it was test flown by a friend in 1930, David taught himself to fly in it. The Kadiak participated in the 1932 Cleveland Air Races. David sold the airplane in 1939. After 16 different owners the Kadiak still exists, and is still in flying condition. David died in 1988.

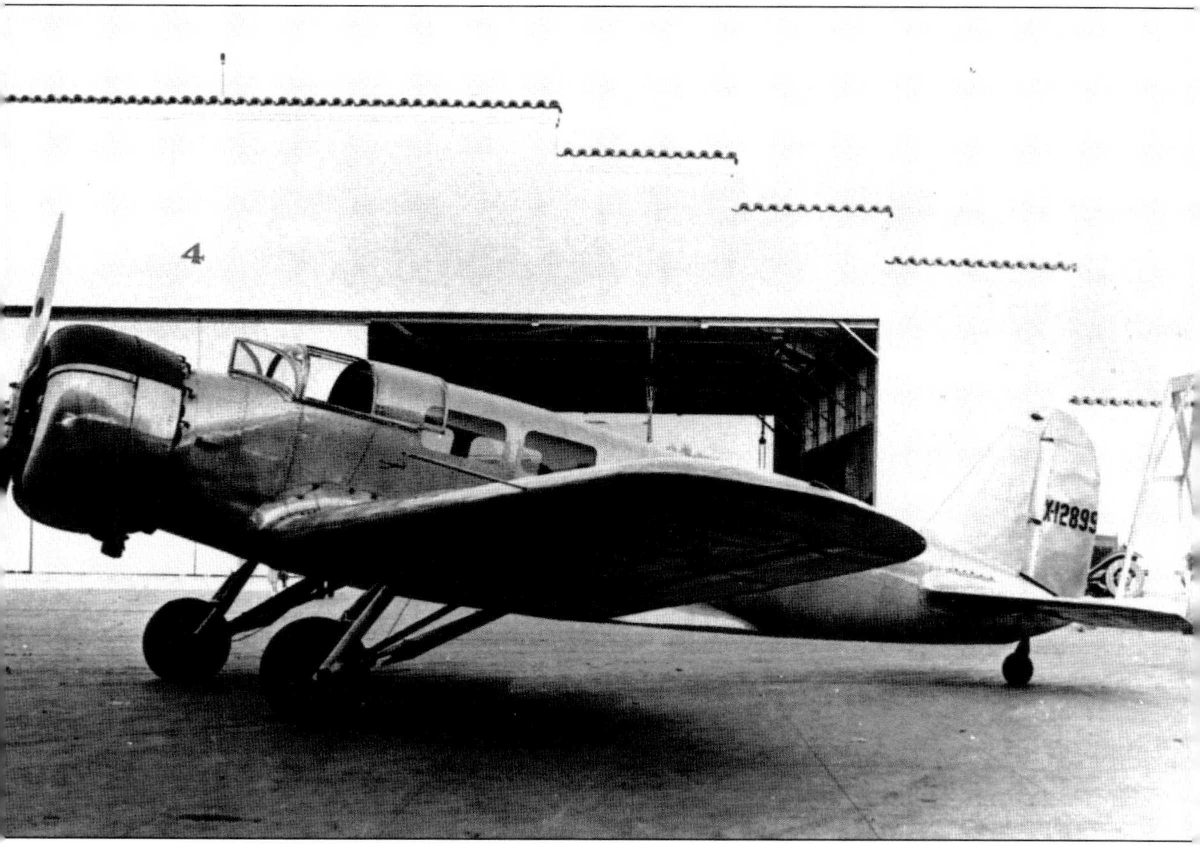

Following the demise of Detroit-Lockheed, test pilot Vance Breese was hired by Air Express, a cargo carrier operating Lockheed Vegas and Orions. Air Express closed in 1932, so Breese decided to build his own cargo-carrying airplane. He obtained financial backing from Charles Dallas, a Detroit automobile dealer and formed the Breese-Dallas Aircraft Company at Detroit City Airport. In six months, Breese and a small crew built this all-metal, six-place monoplane with retractable landing gear. Breese made the first flight in February 1933. In April 1935, he flew the airplane to California and in October 1936, sold it to aviatrix Jacqueline Cochran. Three months later, she sold it to Paul Mantz, and he sold it to a Spanish purchasing agent buying aircraft to use in the Spanish Civil War. On the delivery flight across Mexico to meet a Spanish freighter, the airplane crashed and was destroyed. In later years, Breese worked as an independent test pilot and on October 26, 1940, test flew the prototype of the P-51 Mustang. He died in 1973.

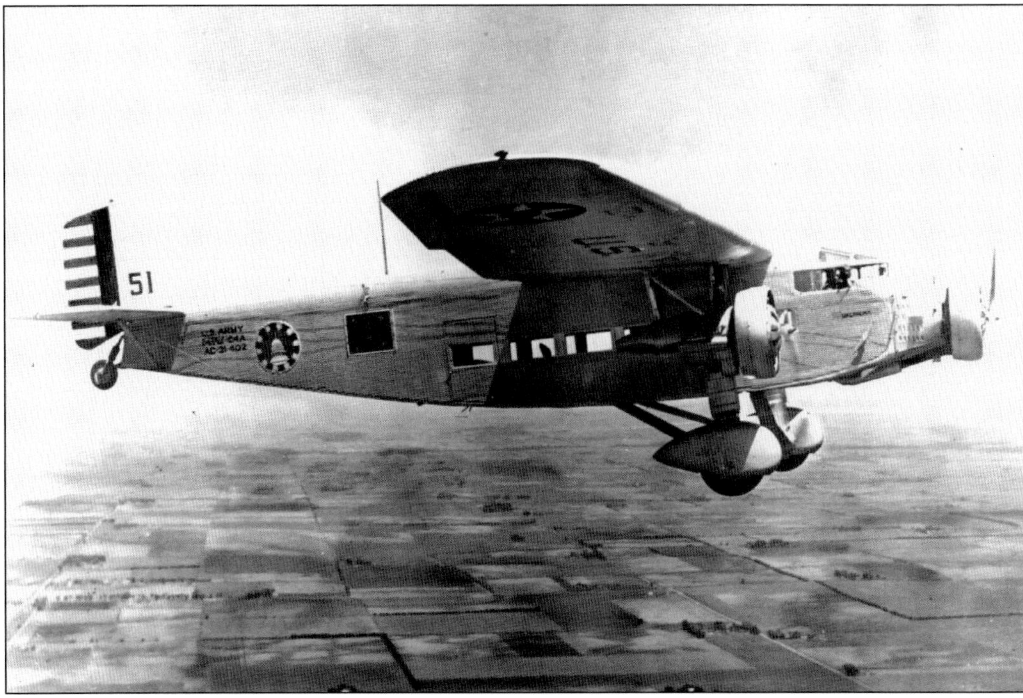

The U.S. military bought 22 Ford Tri-Motors, both 4-AT and 5-AT models. The first 4-AT was delivered to the U.S. Navy in 1927 and designated XJR-1. Two more, designated JR-2, followed. When the 5-AT became available, the U.S. Navy ordered three as JR-3s, two of which went to the U.S. Marine Corps. The U.S. Navy later ordered three more, one designated as RR-4 and two as RR-5s. The U.S. Army Air Corps' first 4-AT was delivered in 1927 as the XC-3. Seven more were ordered as C-9s, plus five 5-ATs designated C-4 and C-4A (seen above). In 1931, Ford built a company funded bomber based on a 5-AT-D airframe (seen below). After testing by the U.S. Army Air Corps, the XB-906 was returned to Ford. On September 19, 1931, it crashed on a test flight, killing Ford's chief pilot, LeRoy Manning, and a mechanic.

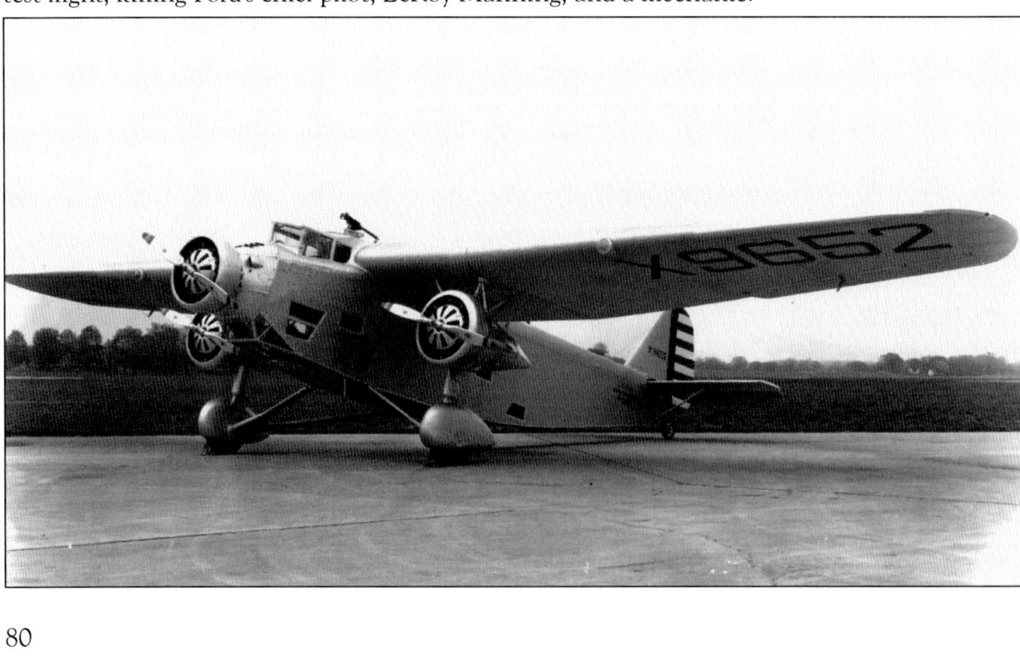

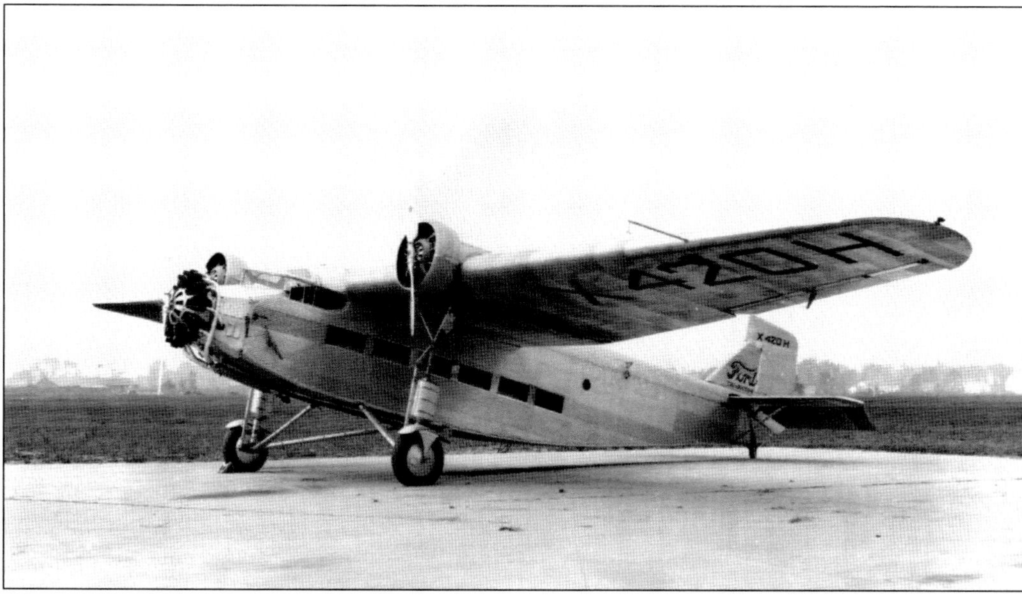

During the production run of 4-AT and 5-AT Tri-Motors, Ford built several experimental versions. This 5-AT (seen above) was built with the two outboard engines mounted in the leading edges of the wings. Performance improvements were minimal so the airplane was converted back to a standard configuration. The Ford 8-AT (seen below) eliminated the outboard engines entirely and replaced the central engine with a single, more powerful power plant. After testing a number of different engines the final choice was a 700-horsepower Wright Cyclone. In 1934, the 8-AT was sold to Pacific Alaska Airways to carry freight and in 1938 was sold to a mining company in Colombia.

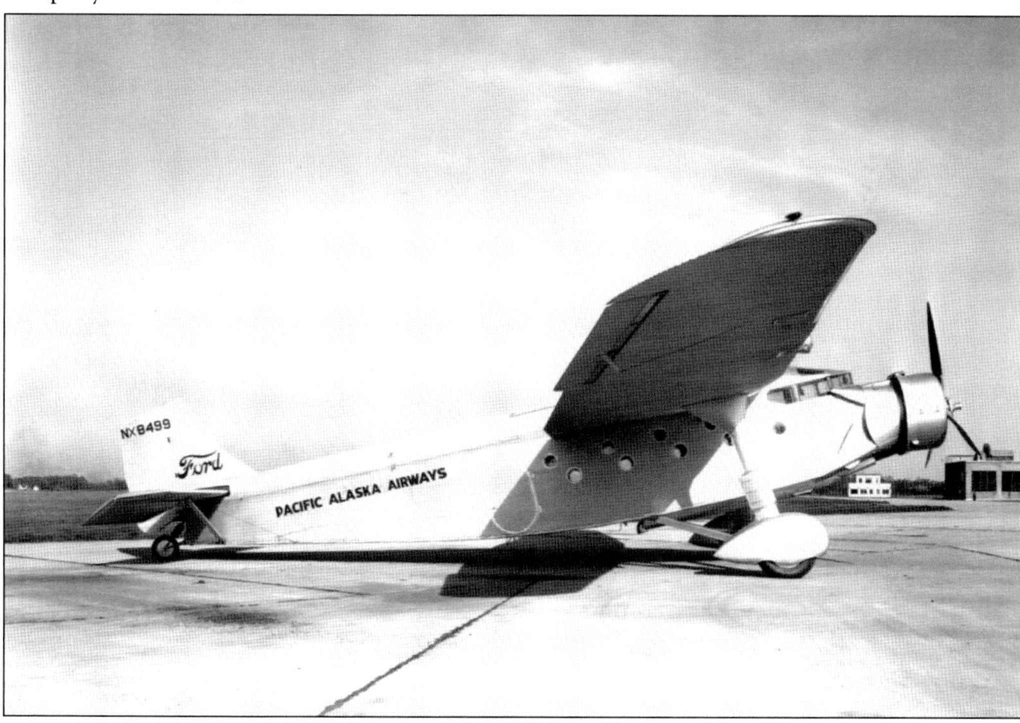

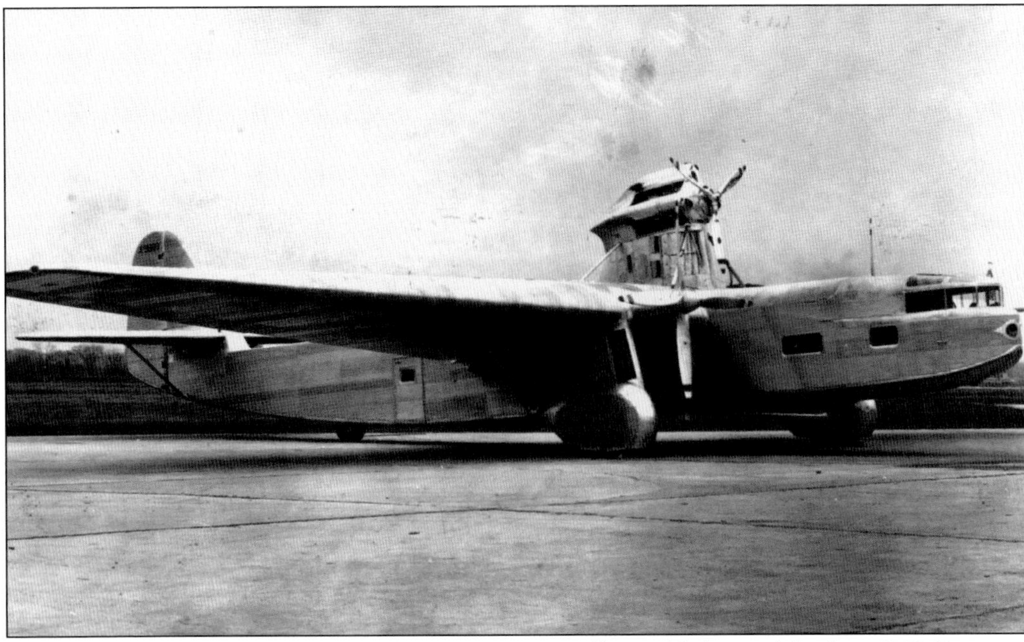

The last Tri-Motor built by Ford was this 32-passenger, all-metal behemoth built in 1932. Designed under the direction of Harold Hicks, the 14-A was powered by three Hispano-Suiza engines, two 715-horsepower V-12s buried in the wings and a single 1,000-horsepower W-18 mounted above the wing. The airplane was far too heavy to fly and after a few taxi tests was scrapped.

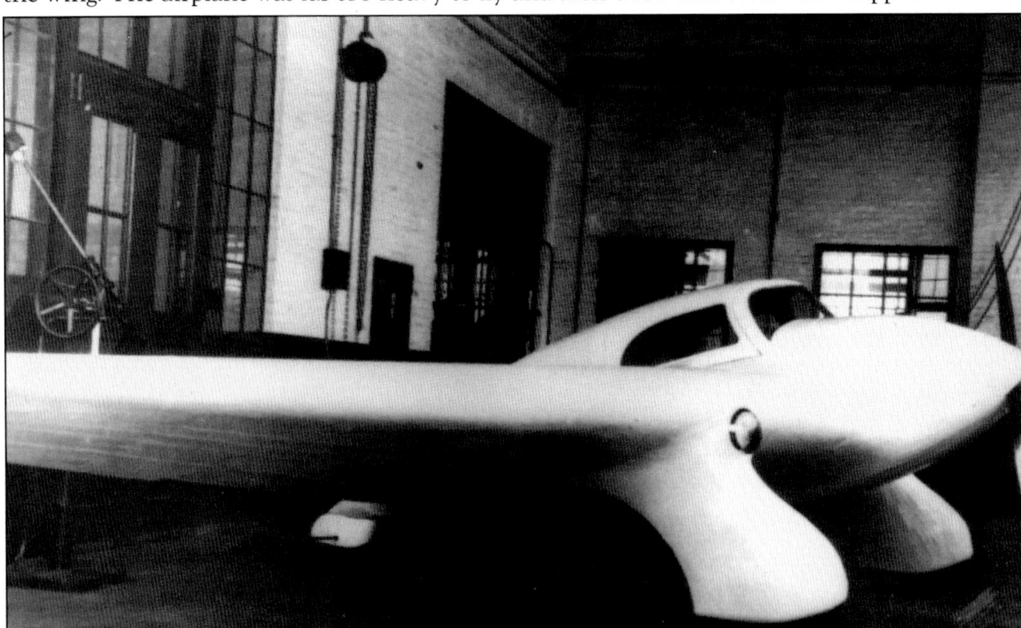

Ford's Aircraft Division's final fling was this two-place, side-by-side personal airplane, the model 15P *Flying Wing*, built in 1935. It had a steel-tube fuselage covered with an aluminum skin and fabric-covered wings. A Ford V-8 engine behind the seats drove a front-mounted propeller via an extension shaft. It reportedly made several short flights before being scrapped. Ford's Aircraft Division closed in April 1936.

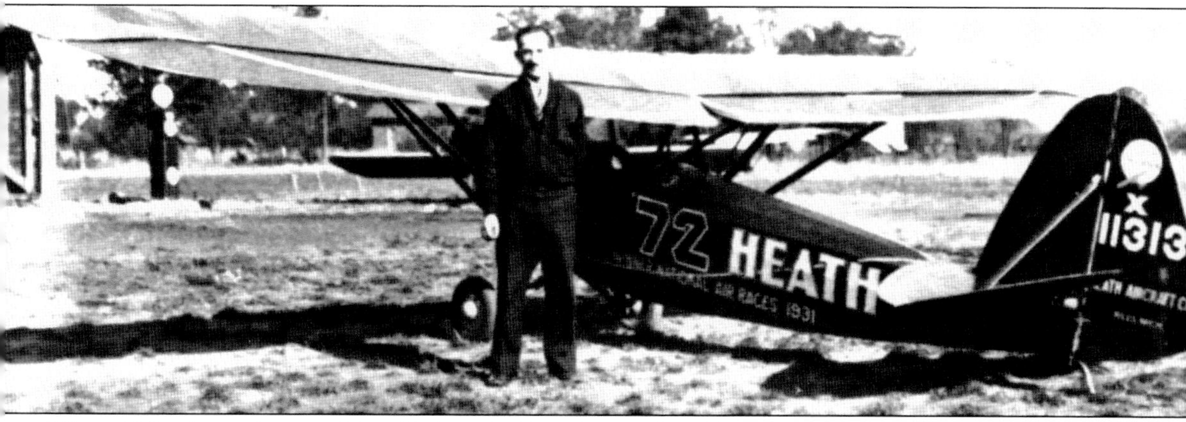

Edward Heath was born in New York in 1888 and moved to Chicago in 1909 to form the Heath Aerial Vehicle Company. Heath built several successful airplanes and operated a flying school and an aircraft supply business. His Heath Parasol, a single-place monoplane, used war-surplus Thomas-Morse Scout wings and sold for $575. The airplane could also be built from a kit or from plans. Heath was killed in 1931 testing a new airplane, and the company was reorganized and moved to Niles, Michigan. Engineers Fred Seiler and Charles Morris designed a new Parasol, the LNB-4, powered by a Heath B-4, a modified motorcycle engine. In 1933, Heath was reorganized as the International Aircraft Corporation and introduced the LNA-40, with a 40-horsepower Continental engine (seen here, with pilot Merrill Lambert), and the CNA-40 Center-Wing. The Niles factory built approximately 9 LNB-4s, 10 LNA-40s, and 5 CNA-40s. In 1937, the company was sold to Howard Anthony of Benton Harbor, Michigan. Anthony sold the aircraft parts and formed Heath Electronics and sold electronic kits.

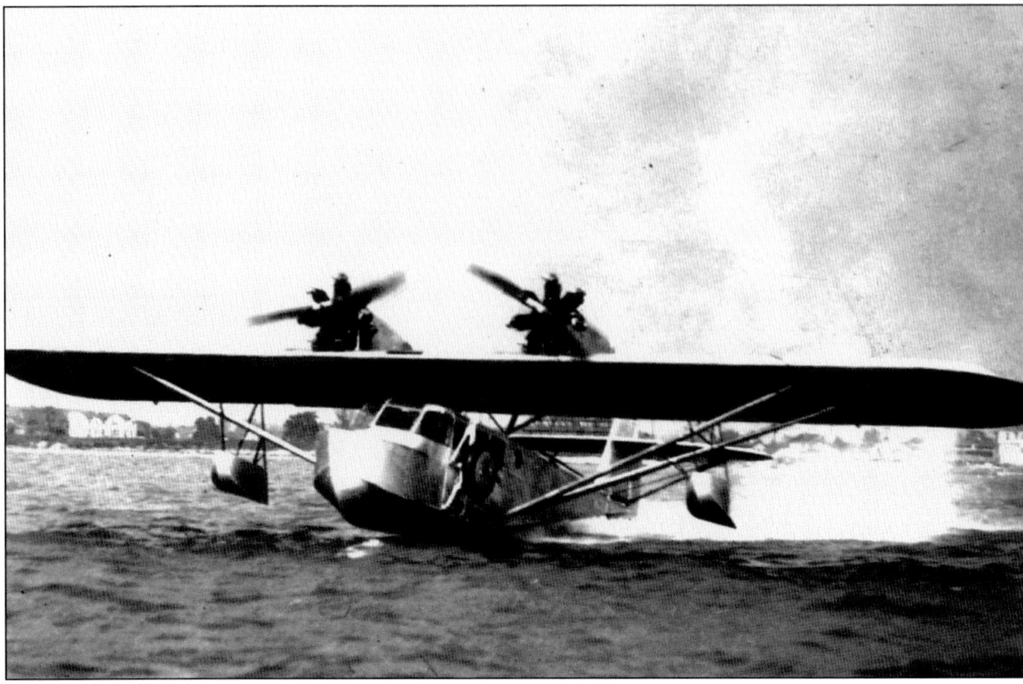

After designing the Eastman Sea Rover, Tom Towle accepted a more challenging project. Henry McCarroll hired Towle to build a twin-engine amphibian for an around-the-world flight to promote Detroit as the "Automobile Capital of the World." The all-wood Towle World Cruiser was first flown in November 1928 (seen above), and McCarroll and pilot George Pond left Detroit in mid-1929. After reaching Brazil, they canceled the remainder of the trip because of concerns about the reliability of the Comet engines. Meanwhile, Towle had obtained backing and formed the Towle Aircraft Company and designed and built an all-metal, twin-engine, six-passenger amphibian. The Towle TA-2 (seen below) was first flown in August 1929. When George Pond was conducting rough water tests, the fuselage broke in two and the TA-2 sank in Lake St. Clair.

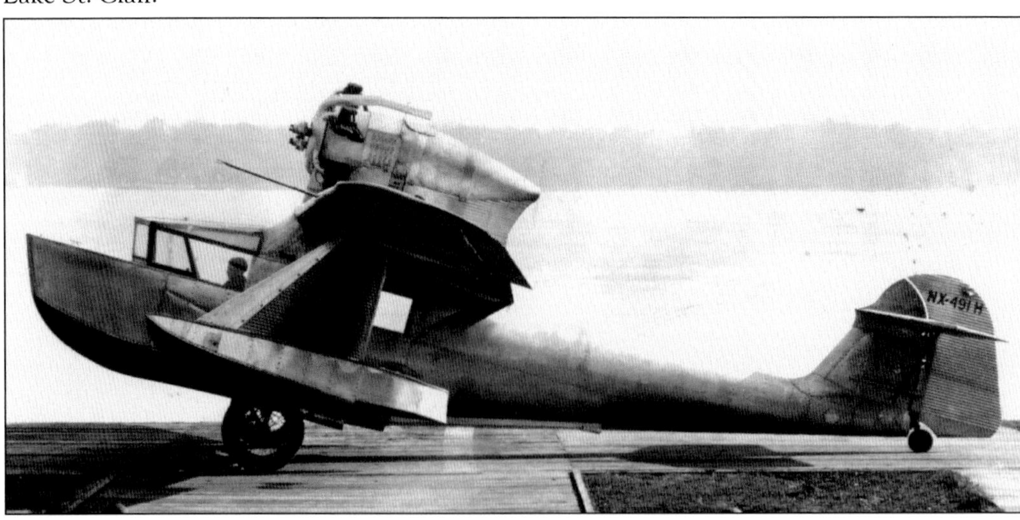

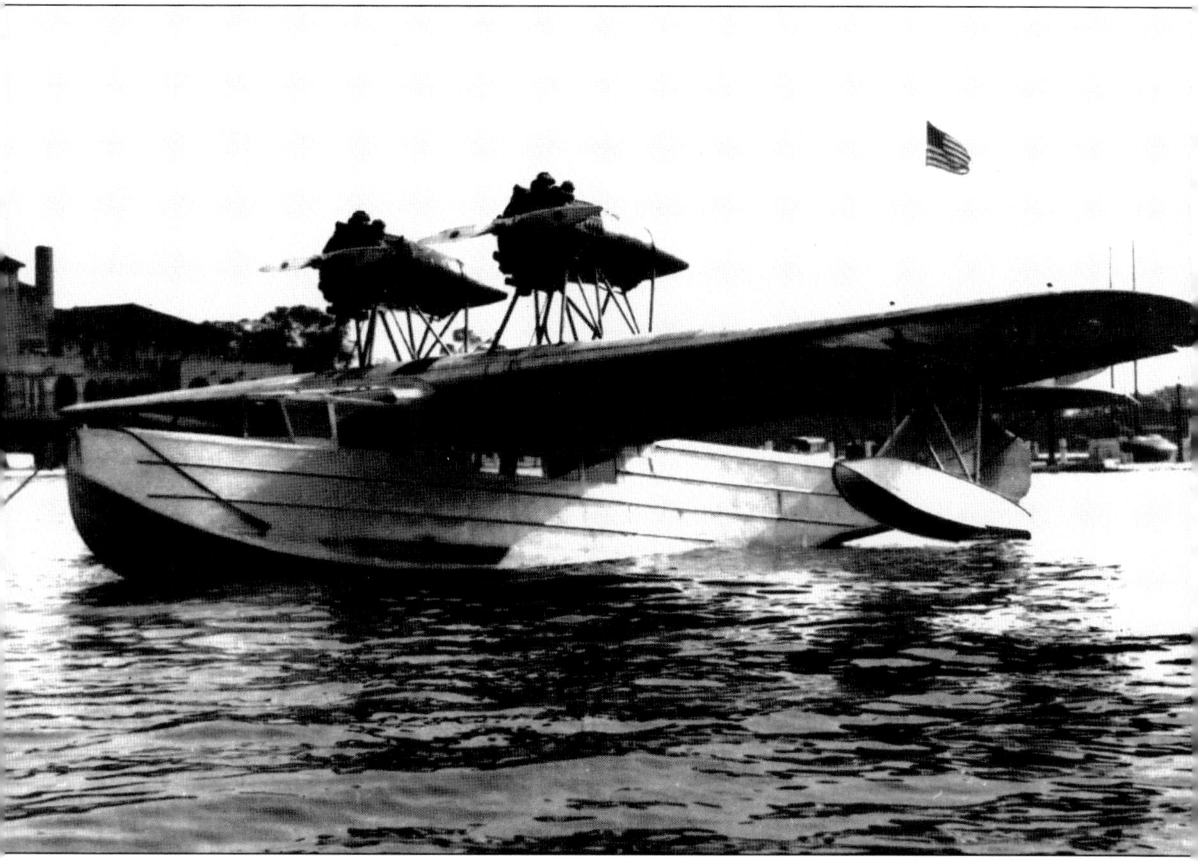

Towle's next airplane was an evolution of the unsuccessful TA-2. The TA-2's wing, which was a scaled-down version of the Ford Tri-Motor's wing, had been salvaged and was still useable. A new six-passenger fuselage was built and assembled with the TA-2's wing and two Packard Diesel engines were mounted on struts above the wing. First flight was made by George Pond in May 1930. The TA-3 was leased to Koehler Airlines for evaluation. In May 1932, a Koehler pilot landed with the wheels in the "down" position and flipped over in shallow water. The airplane was recovered, rebuilt, and sold in July 1932 to a flying service in West Palm Beach, Florida. This was during Prohibition, and the TA-3 was used to haul bootleg liquor from the Bahamas to Florida. The TA-3 was destroyed on Bimini Island during a tropical storm. Towle worked for several aircraft companies until World War II, then became chief engineer of the Hudson Car Company's aircraft division. He retired in 1967 and died in Miami in 1983.

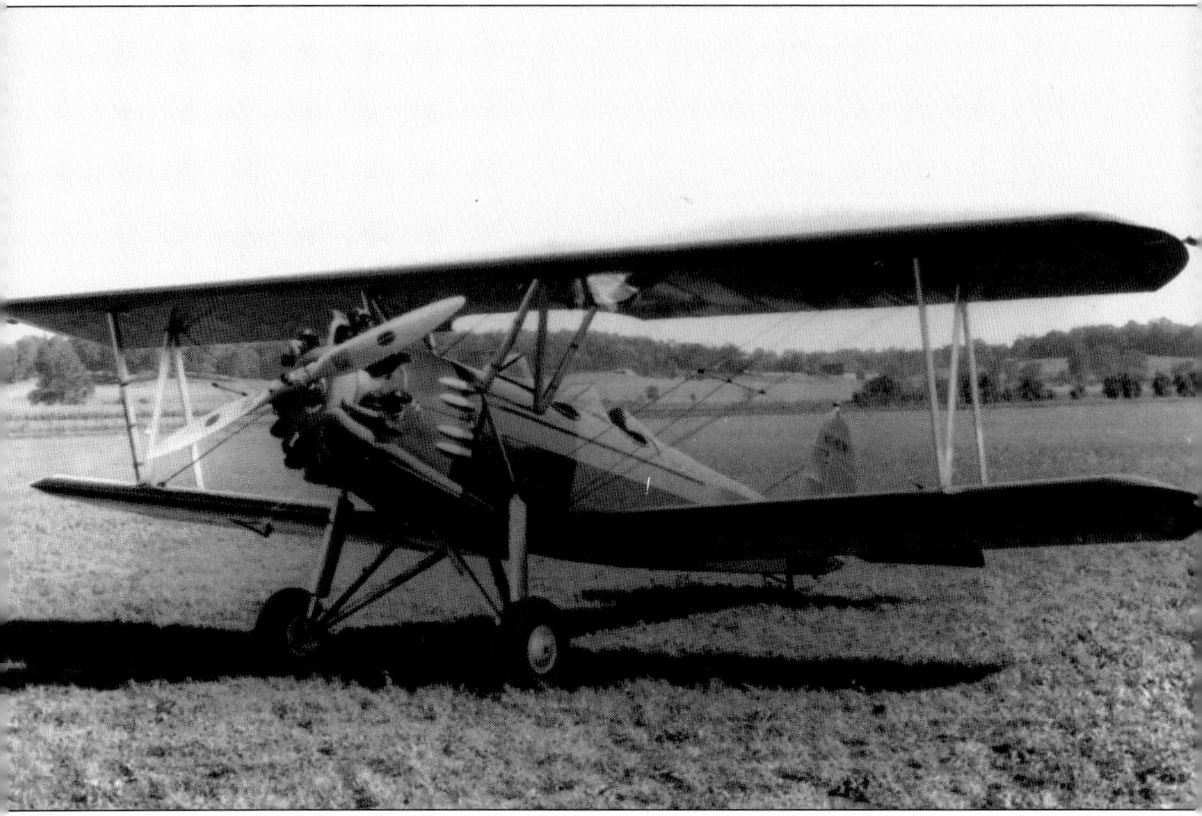

Fred Verville had formed the Verville Aircraft Company in Detroit in 1928 to build the four-place Air Coach, but after the stock market crash, there no longer was a market for an expensive business airplane. Not one to give up easily, in 1930, Verville introduced a sporty, two-place, open-cockpit biplane, the Sportsman. Verville felt there was a market for the Sportsman among wealthy sportsmen-pilots who still had money to spend on a fun-to-fly airplane. He was right, because by the end of 1931, Verville had sold 11 airplanes, plus 4 to the U.S. Army Air Corps. Eventually those markets disappeared, and he was forced to close down. In 1936, Verville was hired by the Bureau of Air Commerce and in 1938 formed an engineering consulting business. After World War II, he worked for the U.S. Navy Bureau of Aeronautics and retired in 1961. Verville died in 1970.

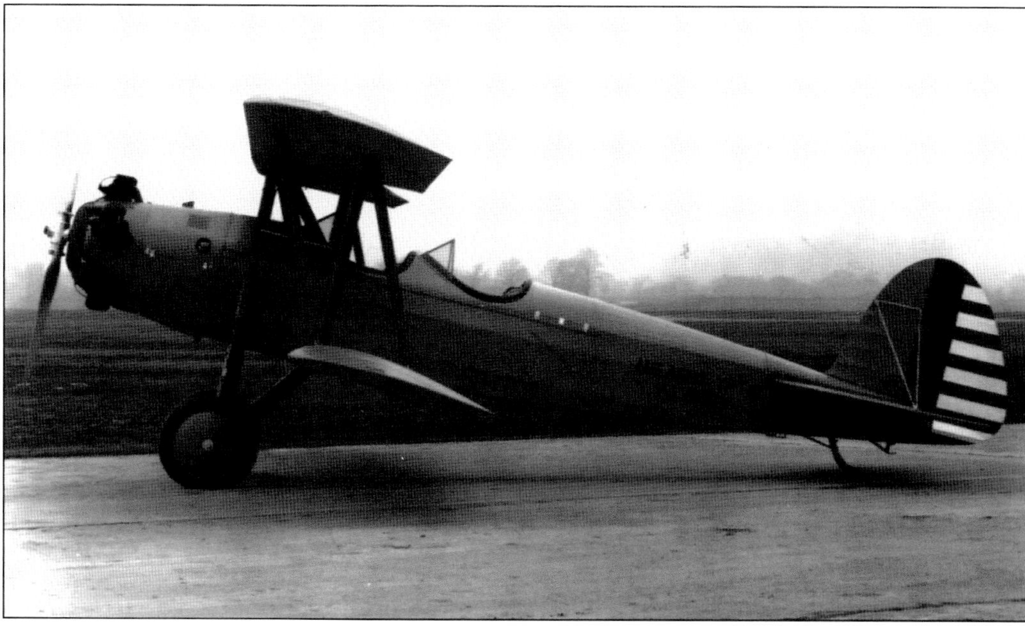

Verville submitted a proposal to the U.S. Army Air Corps for a Primary Trainer versions of the Sportsman. The U.S. Army Air Corps ordered four designated YPT-10 for evaluation (seen above) and tested them with a number of different engines. U.S. Army Air Corps officers felt the YPT-10 would make an excellent trainer, but no production order was received because of military budget restrictions. One Sportsman was built for the National Advisory Committee for Aeronautics (NACA) to test wide-track, long-stroke landing gears (seen below). The airplane was never delivered because on January 31, 1931, it crashed on a test flight when pilot Lou Meister could not recover from a flat spin, bailed out too late, and was killed.

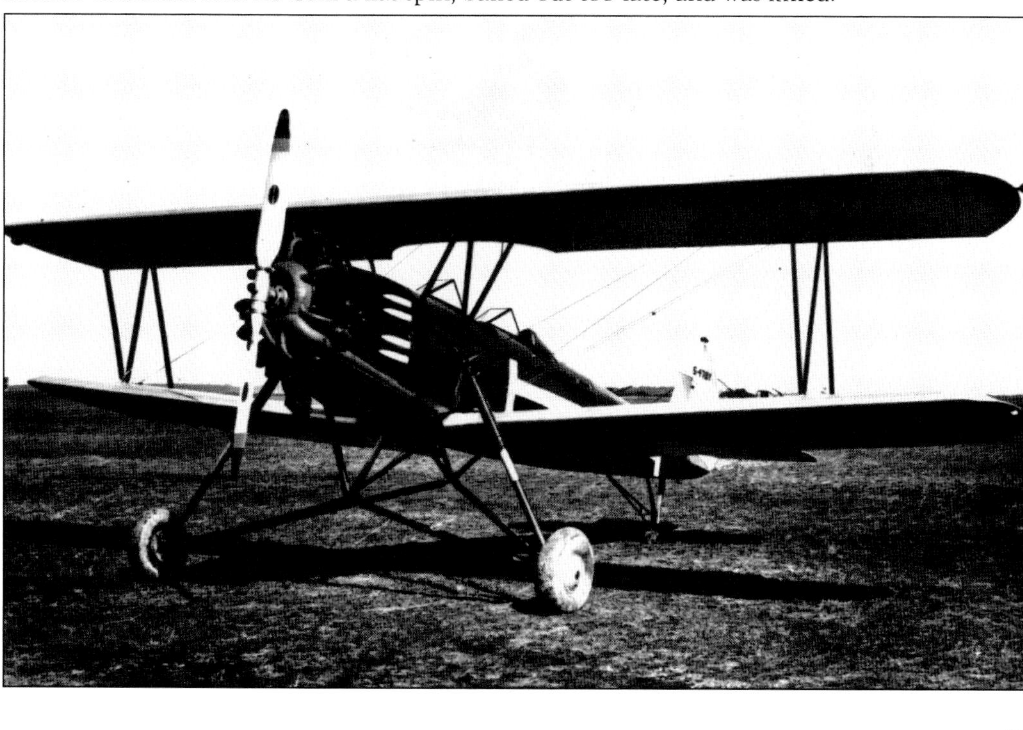

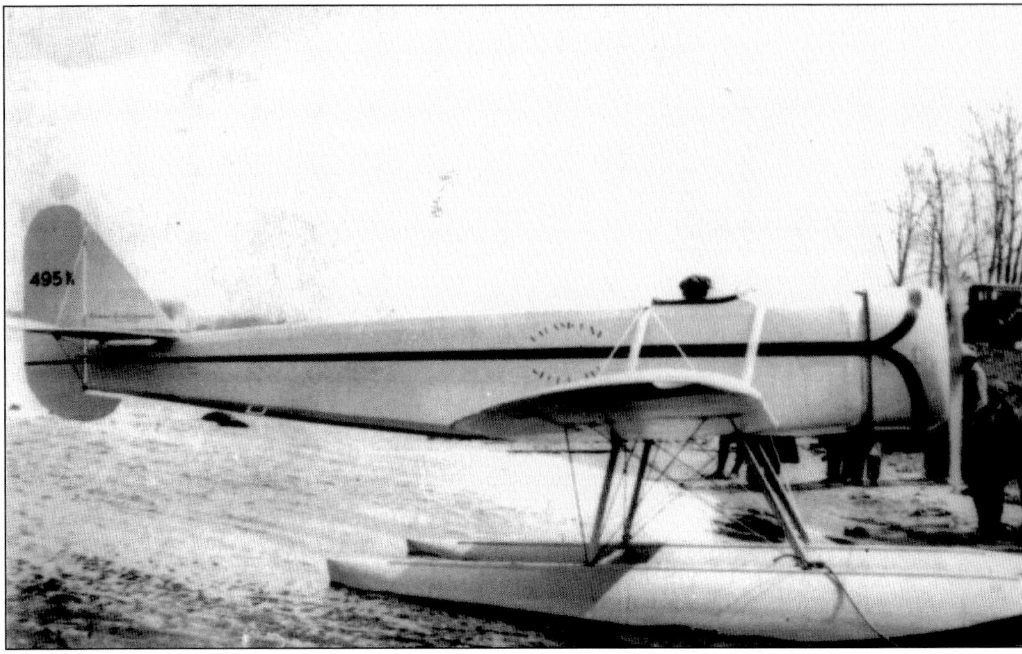

In 1931, Paramount Aircraft of Saginaw was nearing bankruptcy, but president Joe Behse was optimistic. Engineer Ralph Johnson had designed a two-place floatplane, the *Sportster*, that looked like a sure winner (seen above). On May 16, 1931, shortly after its Detroit Aircraft Show debut, Behse was flying the *Sportster* when it crashed into the Saginaw River. That fateful day Joe Behse, his passenger, and Paramount Aircraft all came to a tragic end. During Paramount's final days, cofounder Walter Carr built an OX-5-powered raceplane using leftover parts (seen below) and named her *Miss Saginaw*. He won a number of local and regional air races with the airplane. In 1942, Carr joined the Michigan Aeronautics Commission (MAC), flew military transports during World War II, and retired from MAC in 1962. He died in 1970.

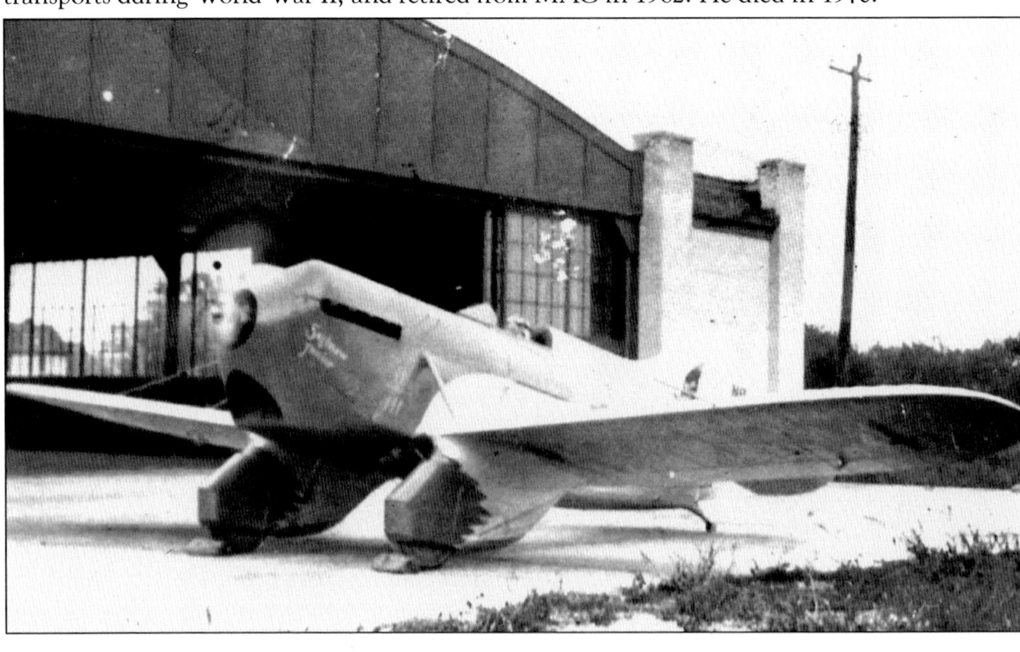

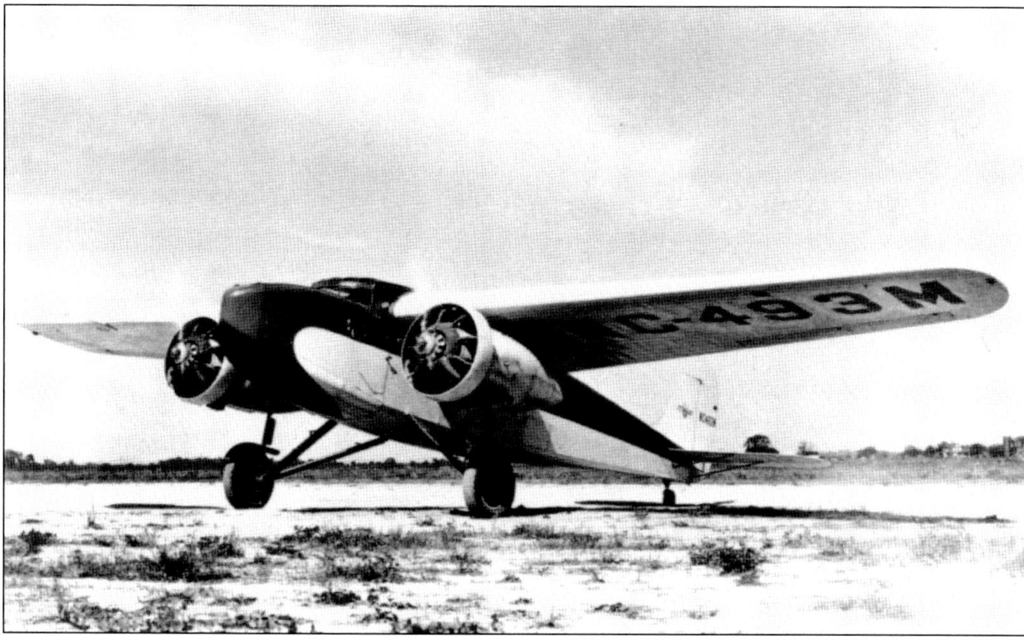

The Stewart Company of Flint built the Stewart M-1 in 1928, but it was not successful. In 1930, president Sid Stewart asked engineer Lionel Kitchen to design a twin-engine, six-place transport, the Stewart M-2, that was test flown by Jack Hunt. Unable to sell the M-2, it sat in a hangar collecting dust. In 1939, the Abrams Aerial Survey Company of Lansing bought the M-2 but found it unsuitable for aerial photography, and it was scrapped.

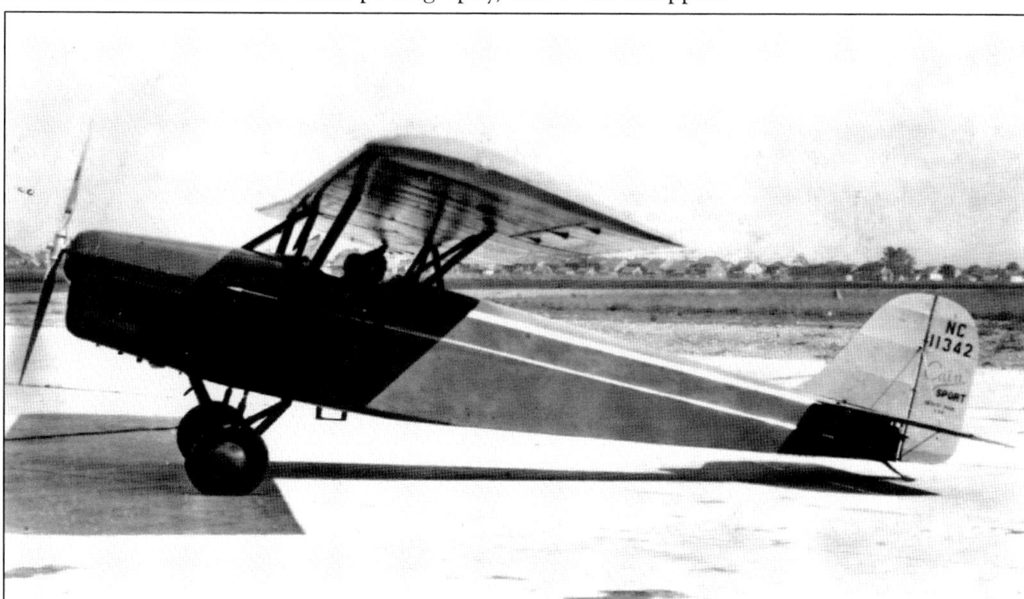

Harry Cain, assisted by engineer Kenneth Ronan, built this Cain CC-14 Sport, a two-seat trainer with side-by-side seating. After exhibiting it at the 1931 Detroit Aircraft Show, it was test flown at Detroit City Airport. In June 1931, the CC-14 was destroyed in a spin accident that killed the test pilot. A second CC-14 was more successful and was flown until 1938, when it was badly damaged in an accident in Minnesota.

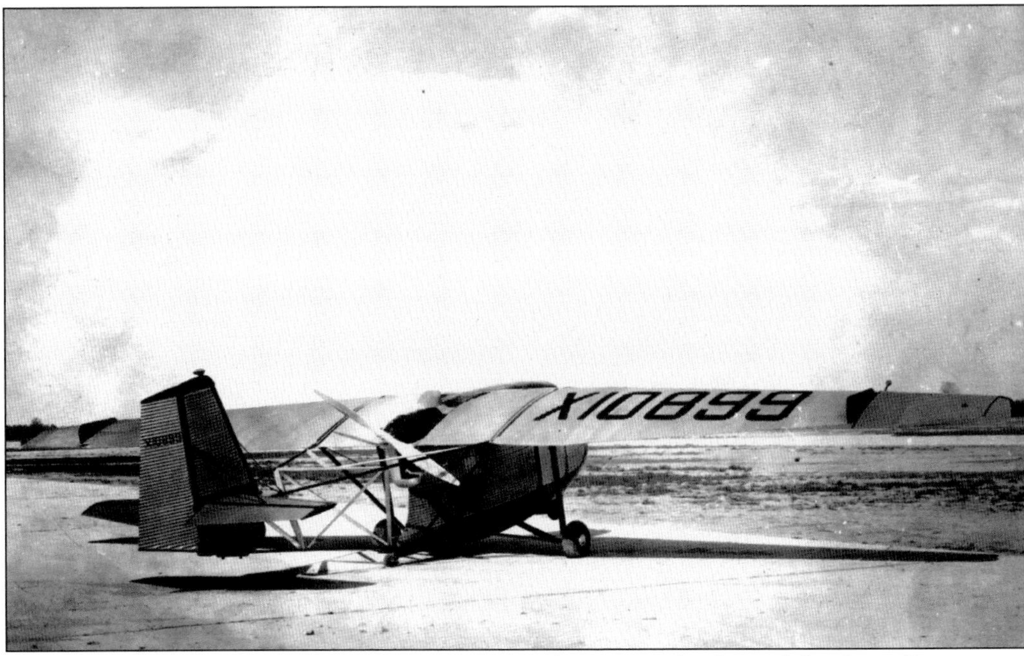

After Bill Stout left Ford Motor Company's Aircraft Division, he formed the Stout Engineering Laboratories in Dearborn. There he built this Stout *Skycar*, a two-place, all-metal personal airplane powered by a 75-horsepower Michigan Rover engine. Stout claimed that anybody could solo the *Skycar* after four hours of dual instruction. The Skycar is presently in storage at the Smithsonian Air and Space Museum awaiting restoration.

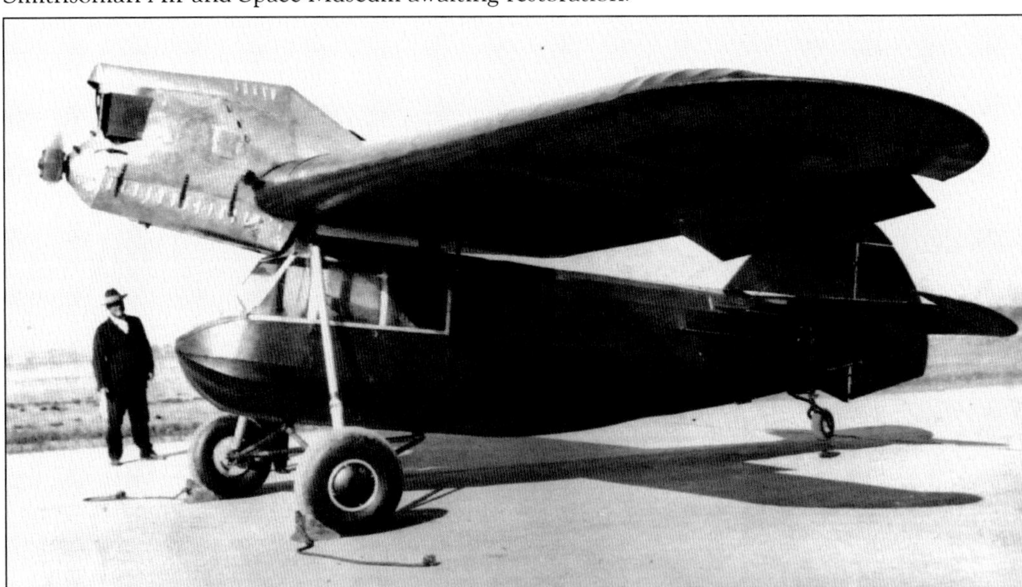

In 1931, the Michigan Airplane Company of Detroit built this ungainly-looking monoplane, the TM-5, also known as the *Taflan Beetle*. The power plant was a 120-horsepower Dayton Bear, a modified, four-cylinder Hall-Scott engine. The man in the photograph was not identified. He may have been the designer who wished to remain anonymous. The airplane was scrapped shortly after its first flight. (Courtesy Wayne County Historical Society.)

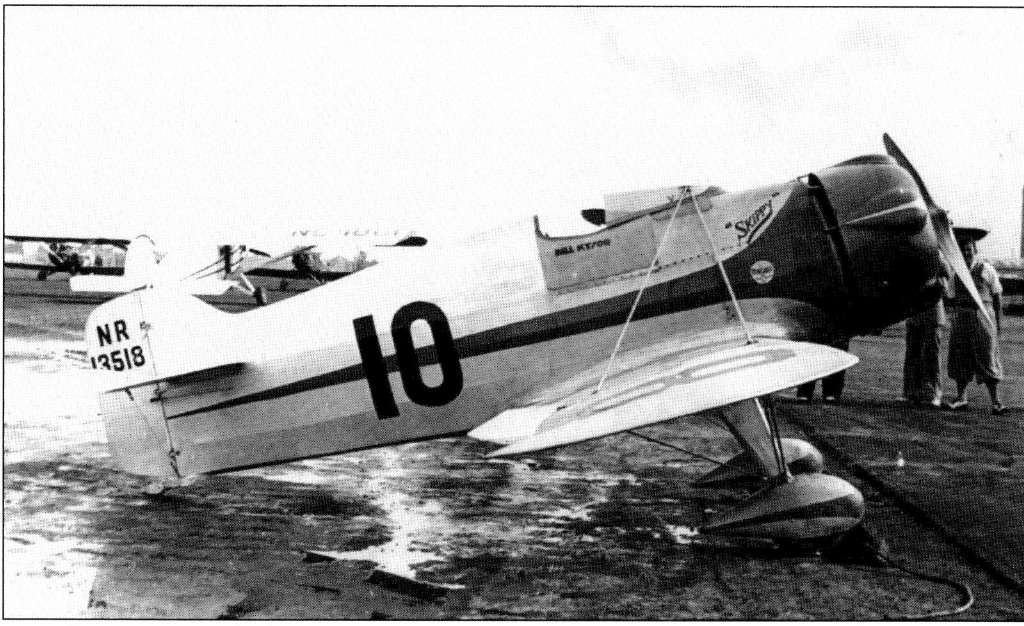

Hans Rasmussen worked as a machinist for Heath Aircraft, Niles, Michigan. In 1933, he built this small racing airplane and named it *Skippy*. Rasmussen also built the airplane's four-cylinder engine. Test flying was by Willis Kysor. *Skippy* was later fitted with a five-cylinder radial engine, designed and built by Rasmussen, and raced in the 1934 National Air Races. Rasmussen returned to his native Denmark in 1935.

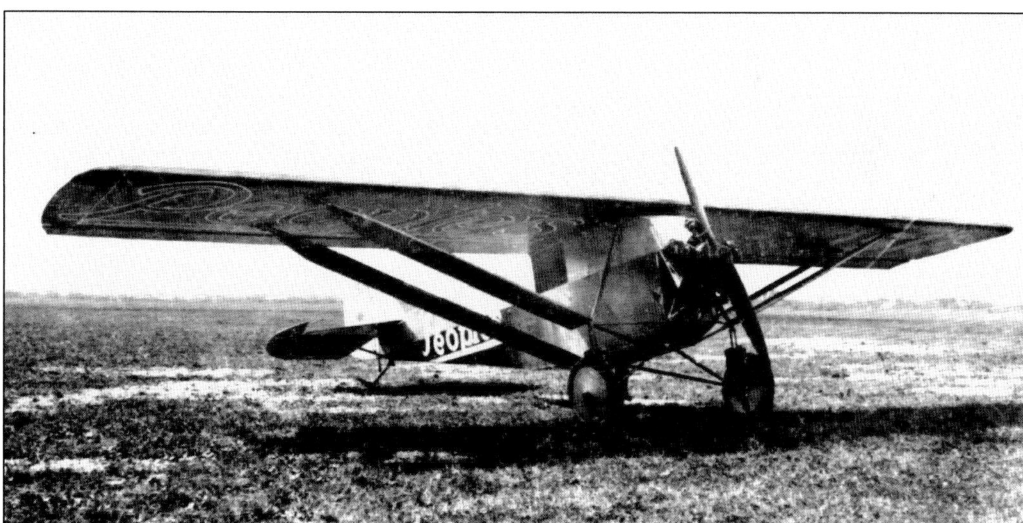

Jack B. Stinson was six years younger than his brother Eddie. In 1922, the brothers formed the Stinson Aircraft Company in Dayton and built a three-place biplane, the *Greyhound,* that later crashed. After Eddie moved to Detroit, Jack followed and formed the Stinson School of Aviation. In 1927, Jack and some students built this cabin monoplane, the JBS *Aircoupe* (seen here), powered by a 60-horsepower Detroit Air Cat engine.

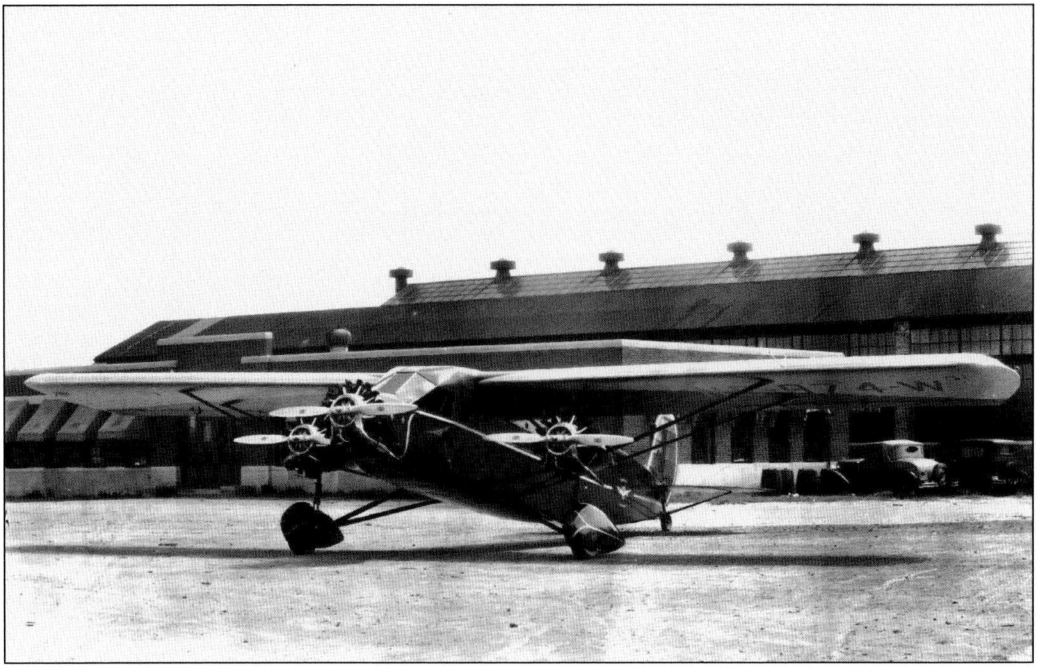

Corman Aircraft was a division of the Cord organization. After Cord bought Stinson, the one and only Corman tri-motor was taken over by Stinson, modified, and displayed at the 1931 Detroit Aircraft Show as the Stinson model SM-6000, also known as the model T. The low price of $25,000 was made possible by using Lycoming engines, built by another Cord division. Fifty-three of these 10- to 12-passenger airliners were built.

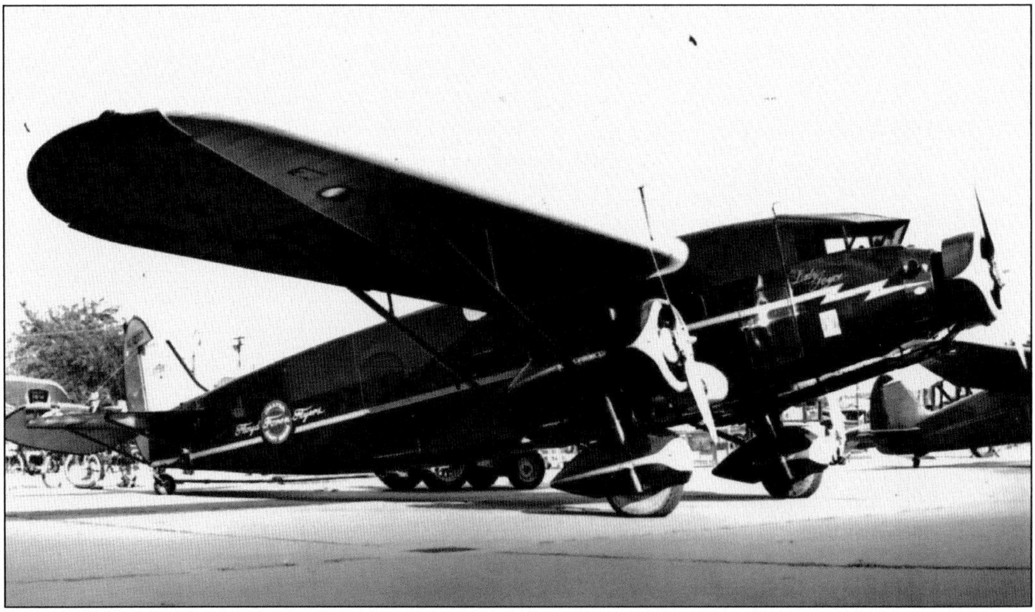

In 1932, Stinson introduced the model U tri-motor, a larger version of the SM-6000. This 12-passenger airliner had short wing stubs to which the outboard engines and the landing gear were attached. Some 24 were built with 18 sold to American Airways, another Cord division. One was bought by newspaper magnate William Randolph Hearst with an eight-seat club interior.

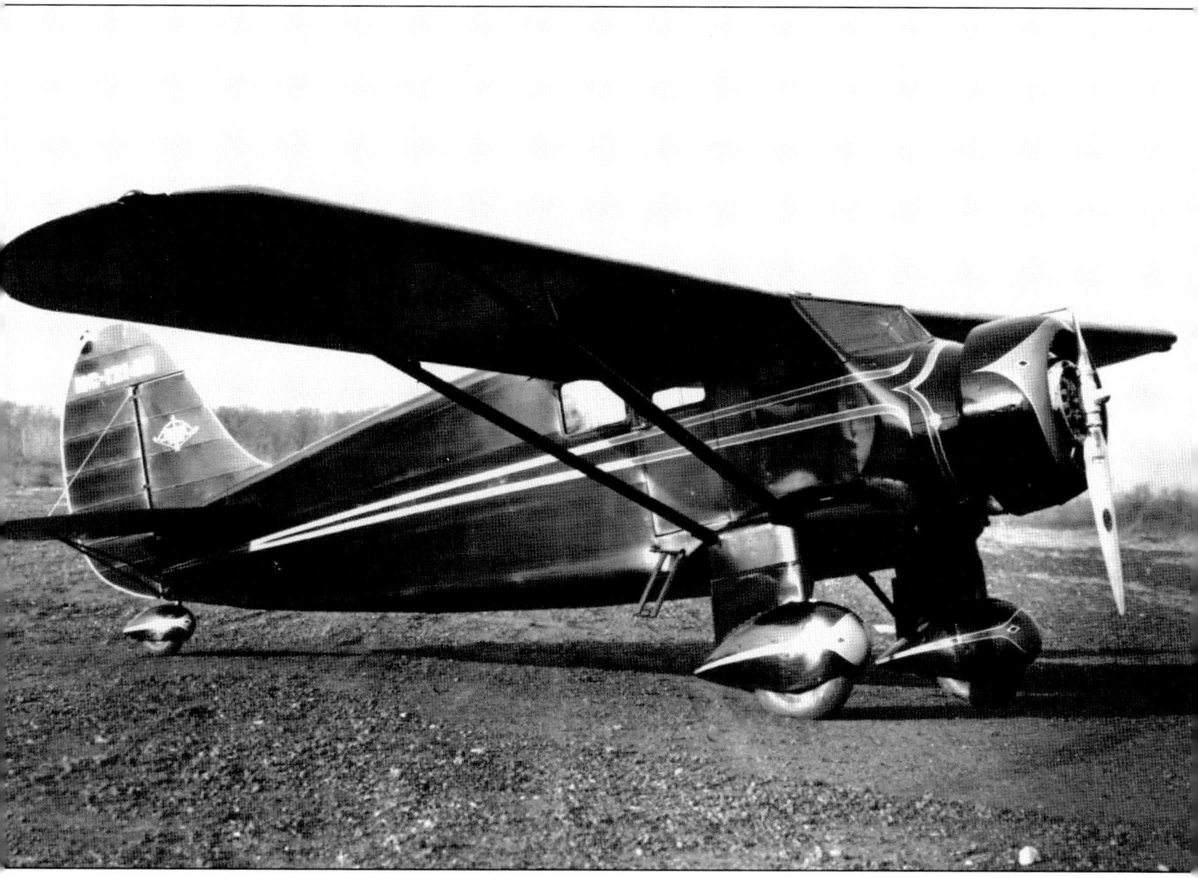

In 1931, engineer Robert Ayer, who had worked for the Granville Brothers on the Gee Bee racers, was hired by Stinson Aircraft. He and Jack Irvine were assigned to design a faster version of the popular Stinson Junior. Starting with a model S, they shortened the fuselage by 30 inches and attached the landing gear to short wing stubs that made it possible to build the airplane with an optional retractable landing gear. On January 25, 1932, after the model R had received ATC No. 457, Eddie Stinson took the prototype on a nationwide sales tour. First stop was Chicago. Stinson spent the day giving demonstration rides, then took up his last three customers near dusk. After takeoff, the airplane ran out of fuel, and Stinson attempted to land on a nearby golf course but hit a flagpole and crashed. The passengers survived, but Stinson died the next day. Eddie Stinson had been the driving force behind the Stinson organization, and his death was a severe setback, but the company survived and eventually sold 39 model Rs. Only five had retractable landing gears.

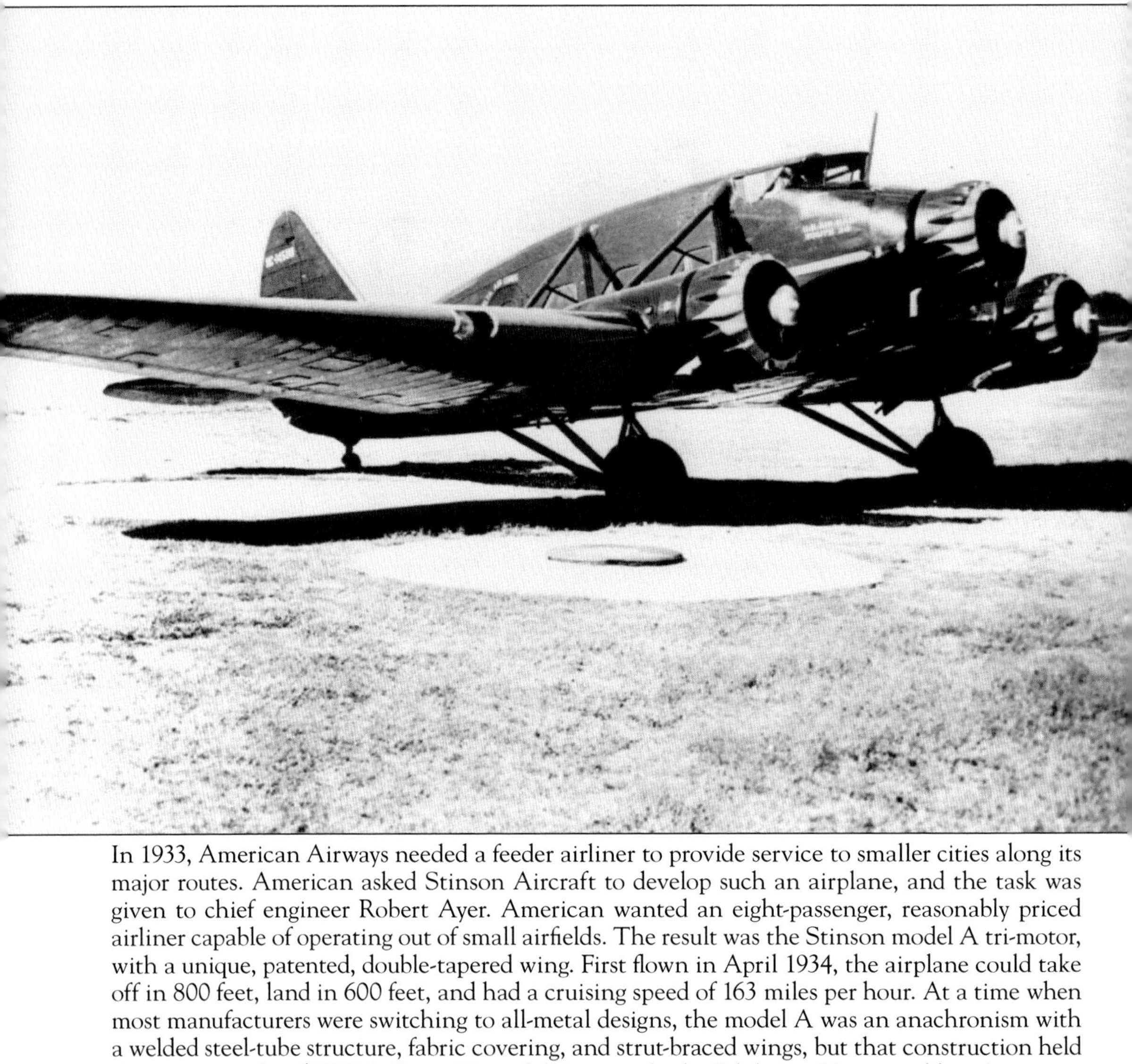

In 1933, American Airways needed a feeder airliner to provide service to smaller cities along its major routes. American asked Stinson Aircraft to develop such an airplane, and the task was given to chief engineer Robert Ayer. American wanted an eight-passenger, reasonably priced airliner capable of operating out of small airfields. The result was the Stinson model A tri-motor, with a unique, patented, double-tapered wing. First flown in April 1934, the airplane could take off in 800 feet, land in 600 feet, and had a cruising speed of 163 miles per hour. At a time when most manufacturers were switching to all-metal designs, the model A was an anachronism with a welded steel-tube structure, fabric covering, and strut-braced wings, but that construction held the price to a low $37,500. Some 31 model As were built. Of those, 16 were sold to American Airways, others to Delta and Central Airlines, and 4 were sent to Australia. One was shipped to China's National Aviation Company in 1936, and Michigan pilot Walter Carr went along to train the Chinese pilots.

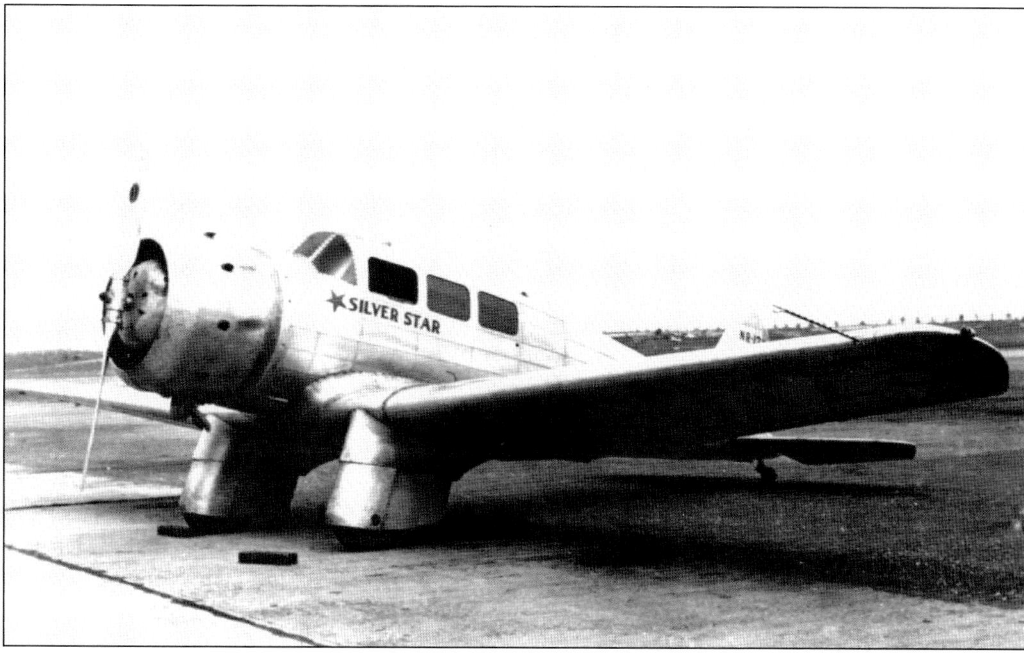

In 1931, Stinson Aircraft built a four-place, twin-engine, twin-boom amphibian (not shown), that was tested on nearby Belleville Lake. Only one was built. Stinson also built a four-place, all-metal experimental monoplane in 1932, the model M *Silver Star* (seen above), powered by a 300-horsepower engine. In 1933, engineer Robert Hall created this two-seat, open-cockpit model O (seen below) by attaching Stinson SR wings and tail surfaces to a new fuselage designed by Hall. Intended for the counterinsurgency role, it was equipped with machine guns and bomb racks. It was built in less than one month and test flown by Hall in May 1933. Nine were sold to military forces in China (three), Honduras (three), Argentina (two), and Brazil (one). Orchestra leader Wayne King bought the prototype, which is still flying.

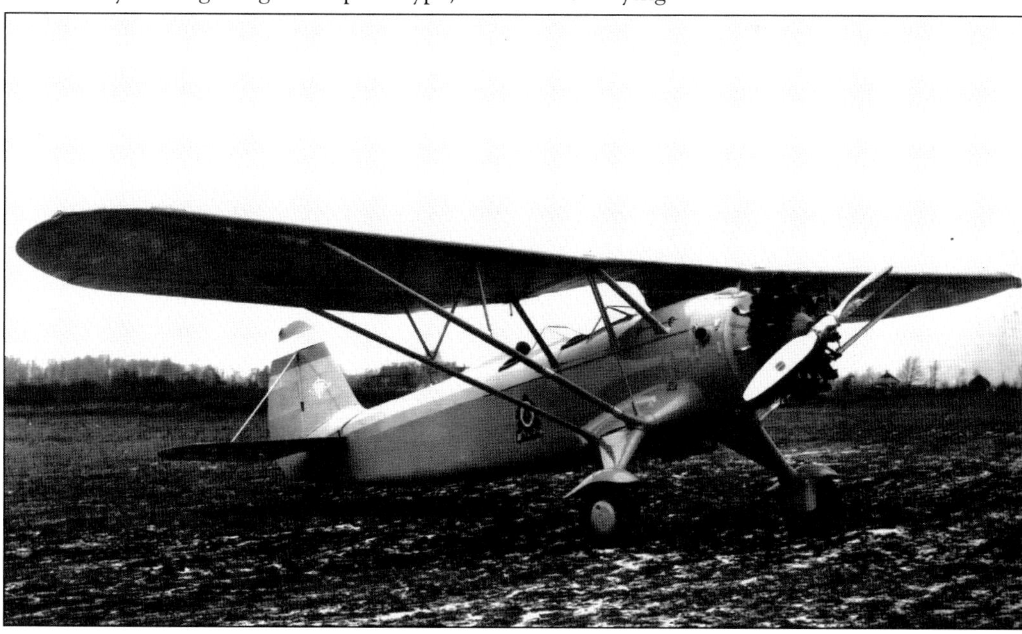

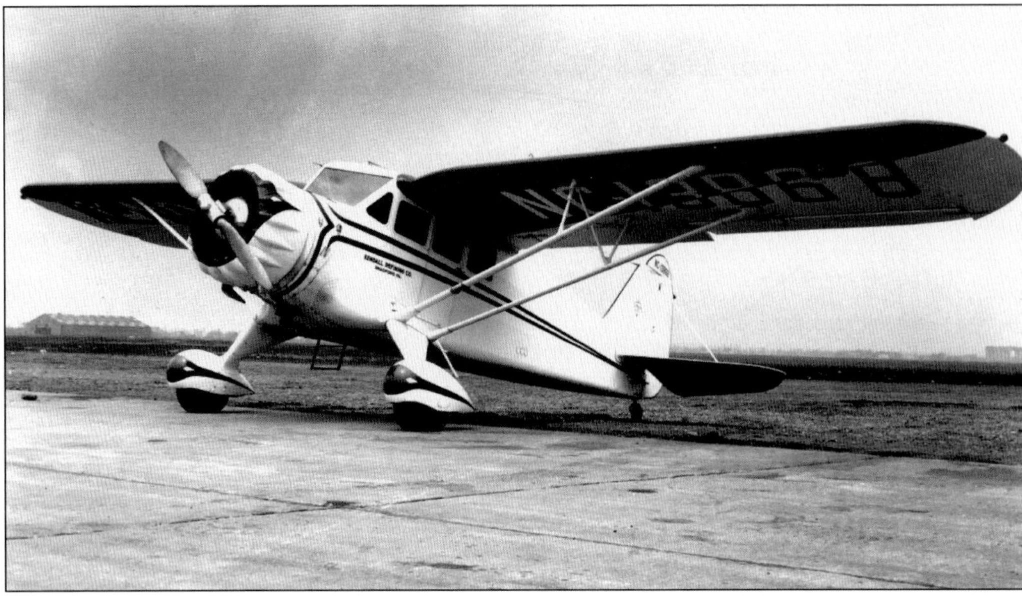

Stinson Aircraft introduced the model SR Reliant in May 1933, priced at $3,995. The Reliant retained most of the basic features of its predecessors, models S and R, but with numerous improvements and a single-strut landing gear (seen above). Models ranged from the SR-1 to the SR-6 and offered a wide choice of options including "Speed Arresters," Stinson's term for wing flaps. Stinson built 287 "Straight Wing" Reliants. Then in 1936, Stinson introduced an elegant new Reliant, the SR-7, that featured a double-tapered wing patterned after the model A tri-motor's wing. The "Gull Wing" Stinsons were built in a variety of models from the SR-7 of 1936 to the last of the series, the SR-10 of 1941. Some models offered Wright or Pratt and Whitney engines instead of Lycomings. Production of the civilian Gull Wing Reliants ended in July 1941 after 488 had been built.

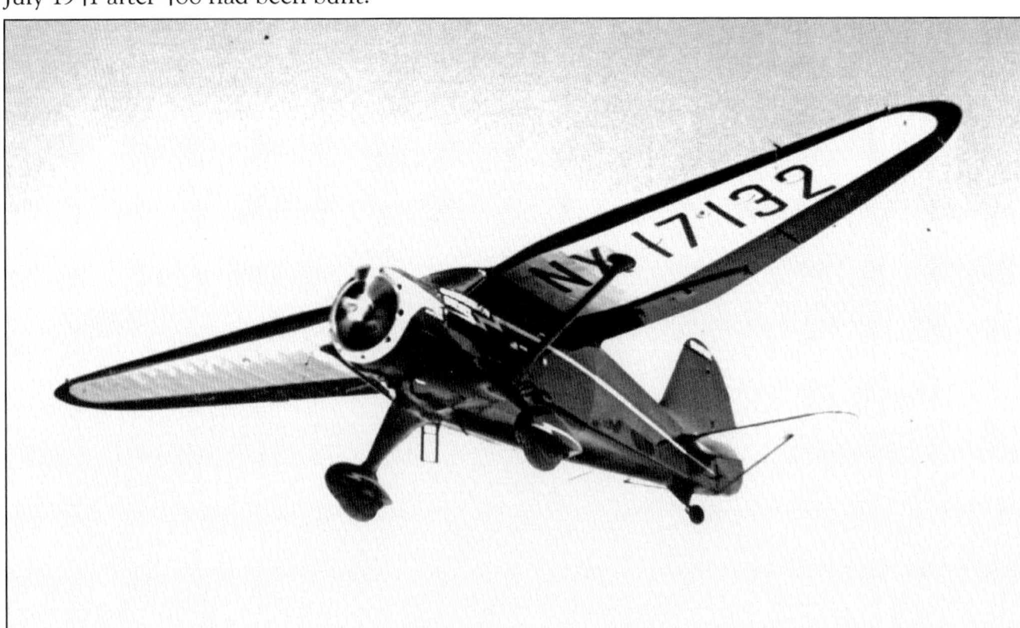

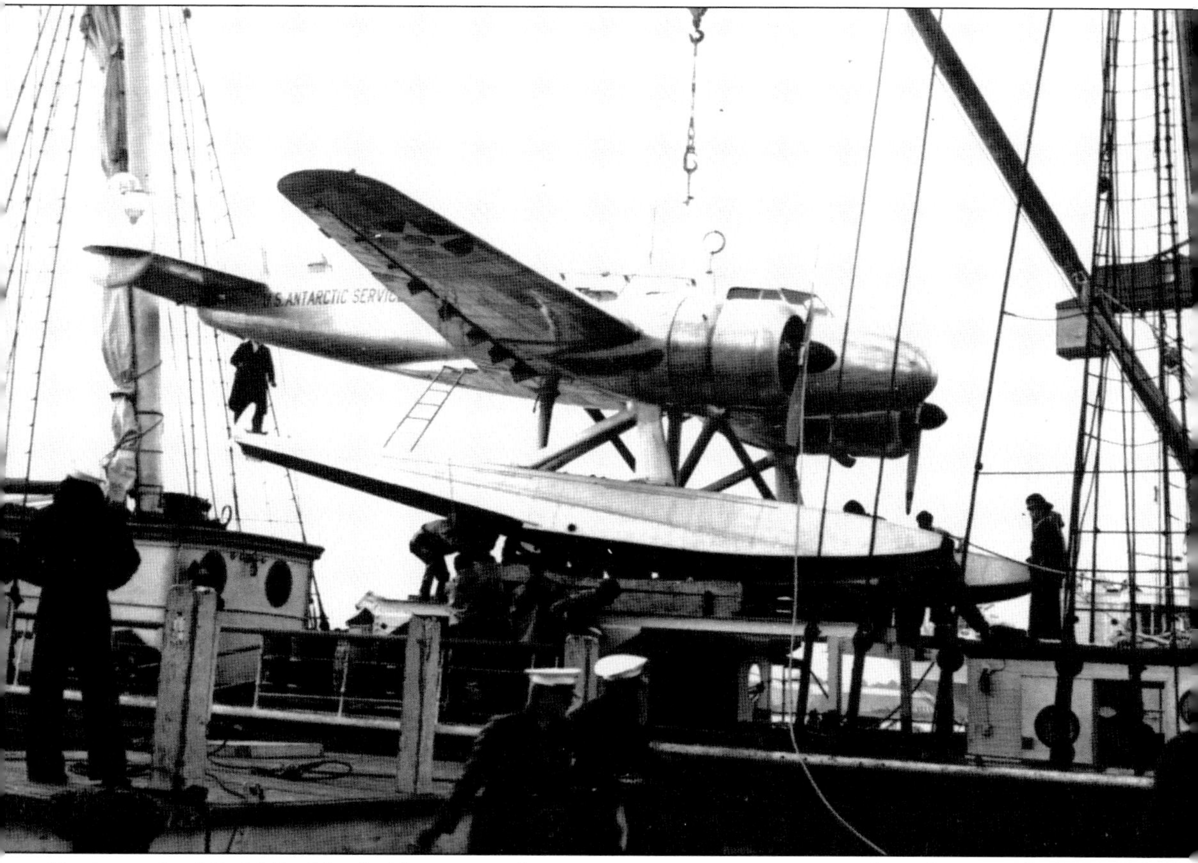

Engineer Archibald Barkley held a patent for a unique wing construction that eliminated ribs by using multiple V-shaped wing spars. A twin-boom monoplane, the Barkley-Warwick, was built in 1931 to test Barkley's invention (not shown). In 1936, Barkley and Harold Grow, an ex-navy commander, formed the Barkley-Grow Aircraft Corporation at Detroit City Airport. Of all-metal construction, the twin-engine, eight-passenger T8P-1 had fixed landing gear making it easier to attach either floats or skis. First flight was in April 1937. Nine were built, and seven were sold to small Canadian airlines carrying passengers and cargo to remote areas in northern Canada. One was sold to Alex Papana for a flight to Rumania that never occurred. That airplane was sold to the Gallino brothers who, in 1940, flew 3,300 miles nonstop from New York to Ecuador, then to Peru where it served in the Peruvian Air Force. Another was purchased by Adm. Richard Byrd in 1939 for his third Antarctic expedition (seen here). In 1940, Barkley-Grow was acquired by Aviation Corporation (Avco) and faded into obscurity.

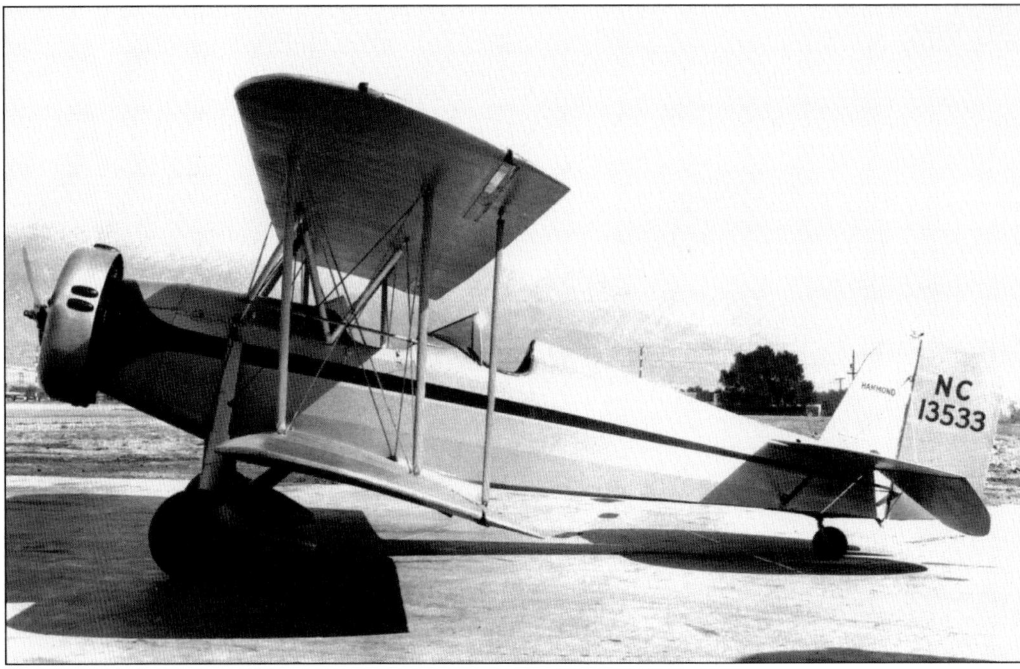

When Detroit Aircraft Corporation declared bankruptcy in 1931, Dean B. Hammond of Ann Arbor, Michigan, purchased the manufacturing rights to the Detroit-Parks P-1H, an open-cockpit biplane. Hammond made some modifications and sold the airplane as the Hammond model 100 Sportster (seen above). Seven were built. When the Bureau of Air Commerce announced a contest to design a two-place, $700 airplane, Hammond built and entered this model Y in the contest. A two-place, twin-boom pusher (seen below), it was of steel-tube construction with fabric covering. Eugene Vidal, head of the bureau, selected the Hammond Y as winner and recruited Lloyd Stearman to assist Hammond in preparing it for production. They formed the Stearman-Hammond Aircraft Corporation in south San Francisco and built a small number of all-metal versions of Hammond's original design.

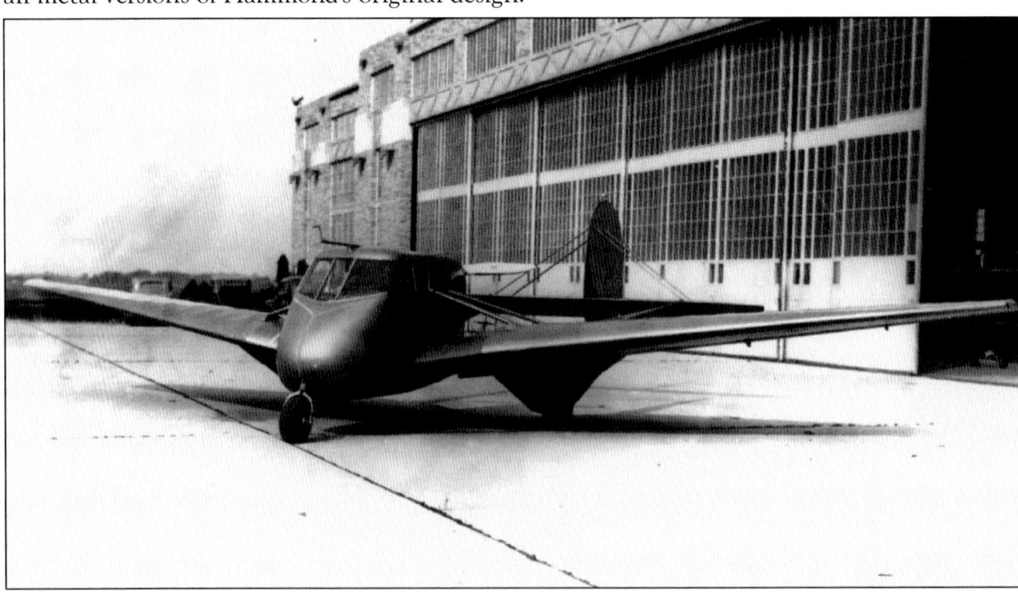

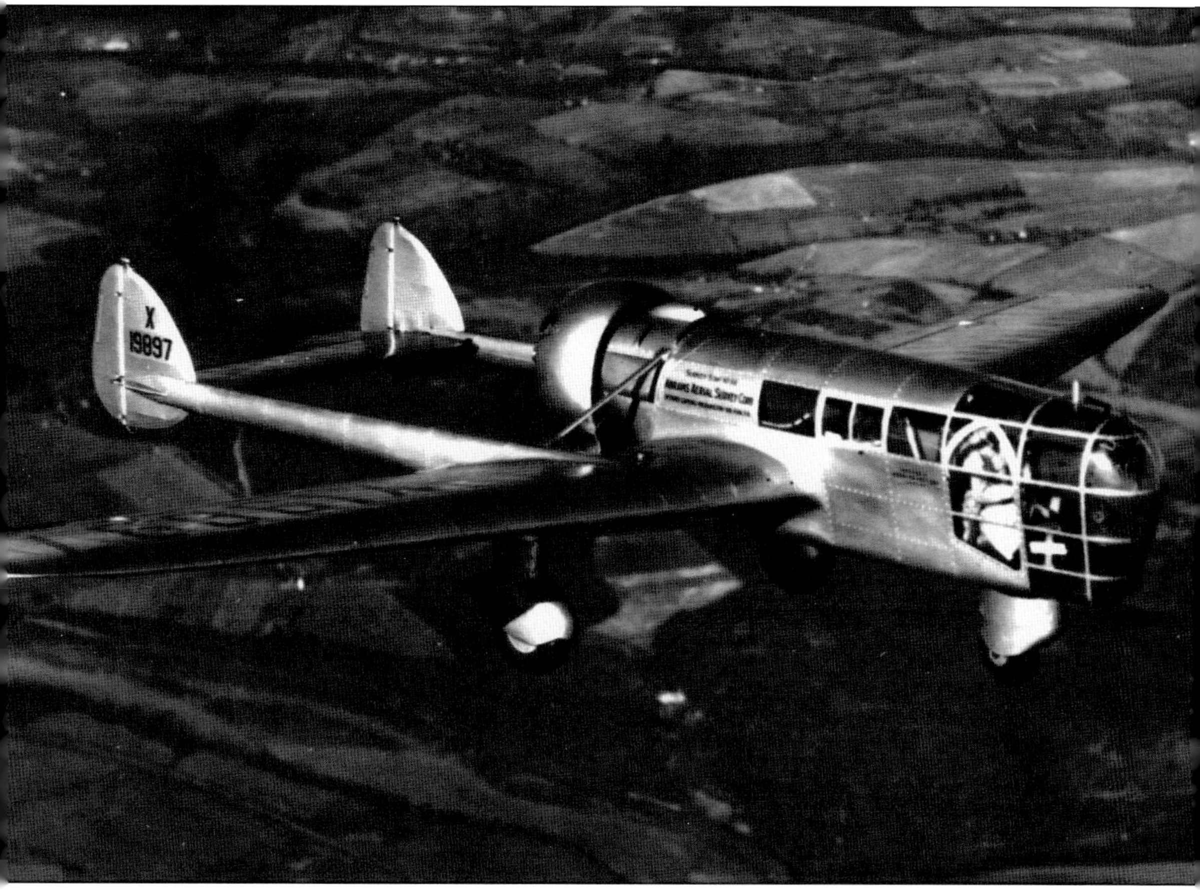

Talbert "Ted" Abrams was born in Tekonsha, Michigan, in 1895. In 1914, he got a job with the Curtiss Aeroplane Company in Buffalo, New York, where he learned to fly. He served in the U.S. Marine Corps during World War I and in 1920 left the service and began barnstorming. Abrams observed that many people were anxious to buy aerial photographs he had taken, so moved to Lansing and formed the Abrams Aerial Survey Company. The business prospered after he had received government contracts for aerial survey photographs. But Abrams was not happy with the ex-military aircraft used for aerial photography, so in 1937, he formed the Abrams Aircraft Corporation to build an airplane to suit his needs. He and engineer Kenneth Ronan designed the Abrams P-1 *Explorer* that was built in Marshall, Michigan. The first flight was in November 1937 by Walter Carr. During World War II, Abrams's company built special aerial cameras and mapping equipment for the U.S. Army Air Corps. After the war, the *Explorer* was donated to the Smithsonian Air and Space Museum.

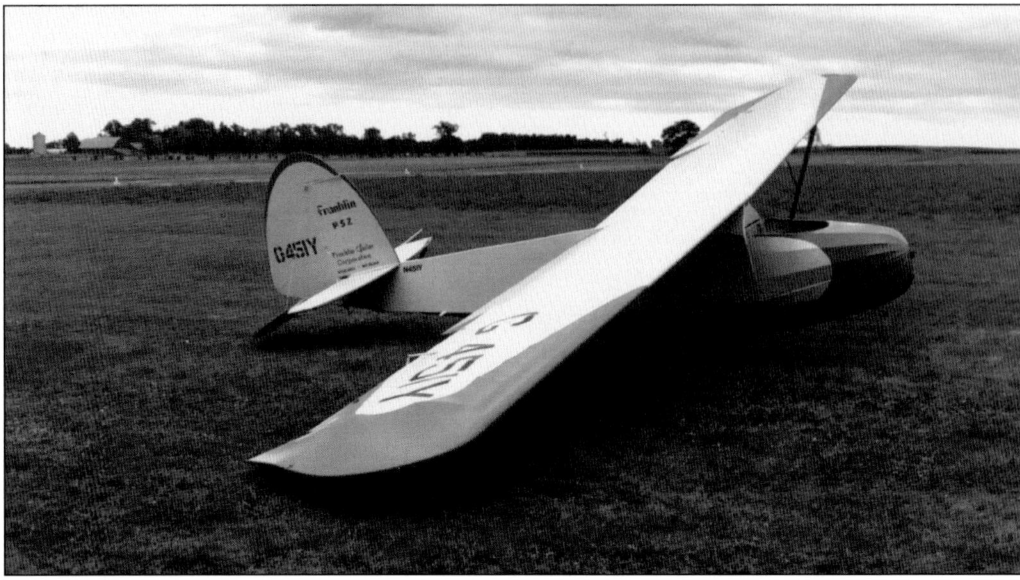

In 1927, University of Michigan students formed a glider club under guidance from Prof. Roswell Franklin and built several primary-type gliders. Professor Franklin saw a need for an advanced-type so in 1928, he built a primary-secondary glider, the PS-2. He then formed the Franklin Glider Corporation to manufacture it (seen above). Approximately 55 Franklin gliders were built. In 1929, Capt. Frank Hawks, a popular racing pilot, won a spot-landing contest in a Franklin glider. Hawks saw gliders as a way to attract youngsters to aviation and conceived the idea of a cross-country glider tow. Franklin built a special glider, the *Eaglet* (seen below, with Franklin on the left and Hawks in the *Eaglet*), and on March 30, 1930, Hawks left San Diego on tow behind a Waco biplane. Eight days and 20 stops later, he landed in New York City. Hawks had been towed 2,800 miles and had given talks promoting aviation at each stop.

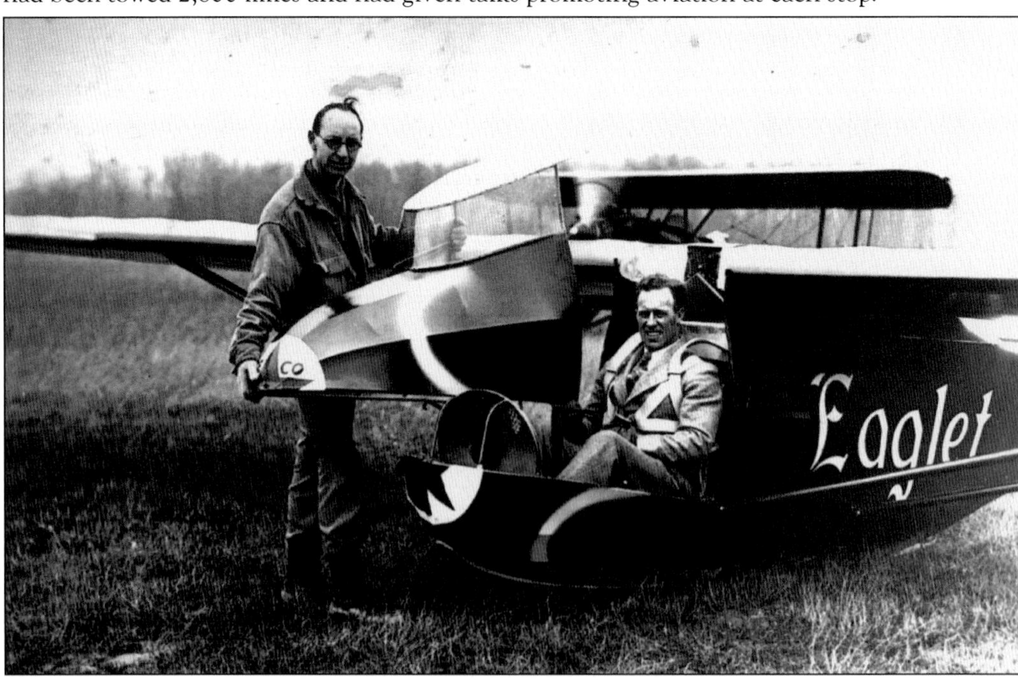

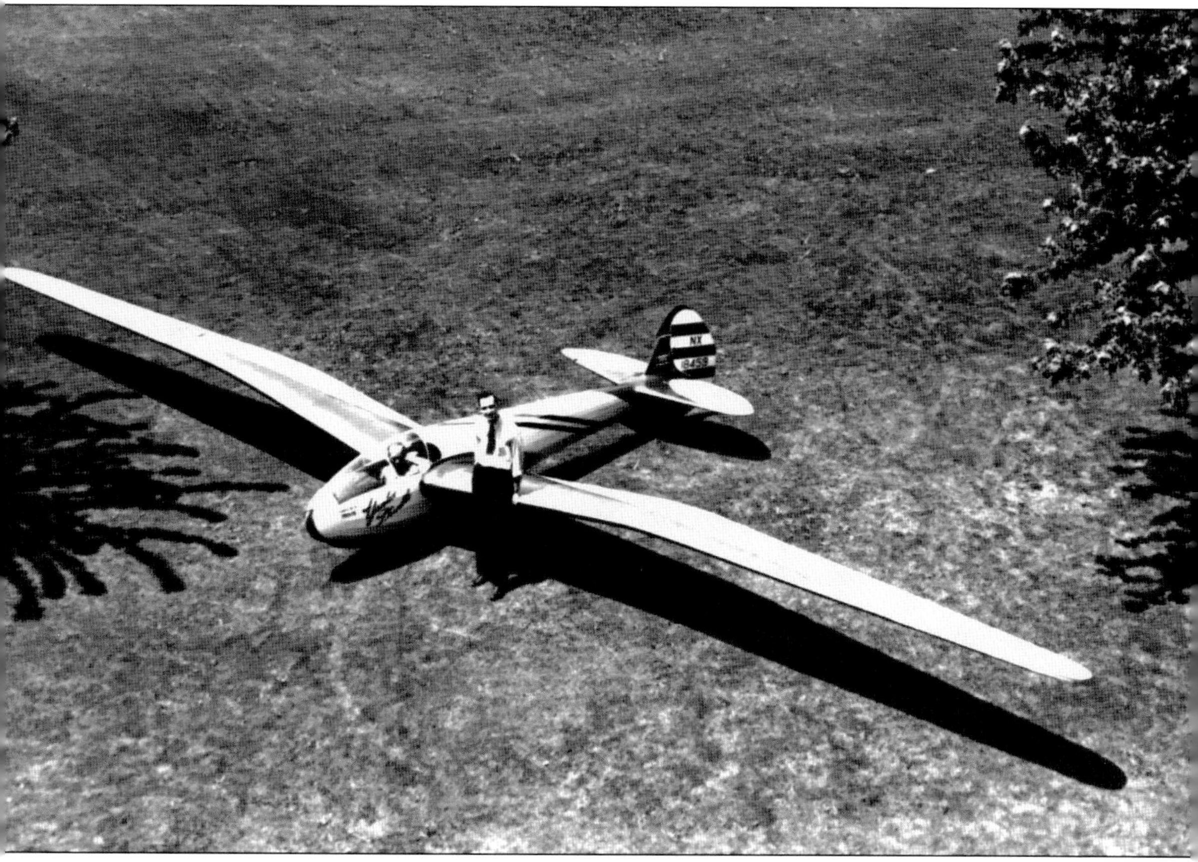

There was renewed interest in glider flying in the 1920s as news of German glider developments reached this country. Many college students formed glider clubs and began building and flying primary-type gliders. In 1928, Edward Evans, a prominent Detroiter, formed the National Glider Association to coordinate those activities. As a result of his efforts, many individuals in the Detroit area were building secondary-type gliders and new glider manufacturers were formed including Gliders Incorporated, Stone Aircraft, Frankfort Gliders, Midwest Gliders, and Franklin Gliders. Lawrence Institute of Technology (LIT) in Highland Park, a Detroit suburb, had become a center for glider activity in Michigan. Aeronautical engineering student Jack Laister formed the LIT glider club using his own two-place glider for training. In 1938, Laister designed a graceful sailplane, the *Yankee Doodle* (seen here). In 1939, LIT was invited to enter that sailplane in a contest in Paris. Laister had graduated by then, so Dick Randolph flew *Yankee Doodle* to a fifth place finish. In 1941, Laister formed the Laister-Kauffman Airplane Company in St. Louis and built a large number of two-place training gliders, the TG-3A, for the U.S. Army Air Corps. The TG-3A was a military version of the *Yankee Doodle*.

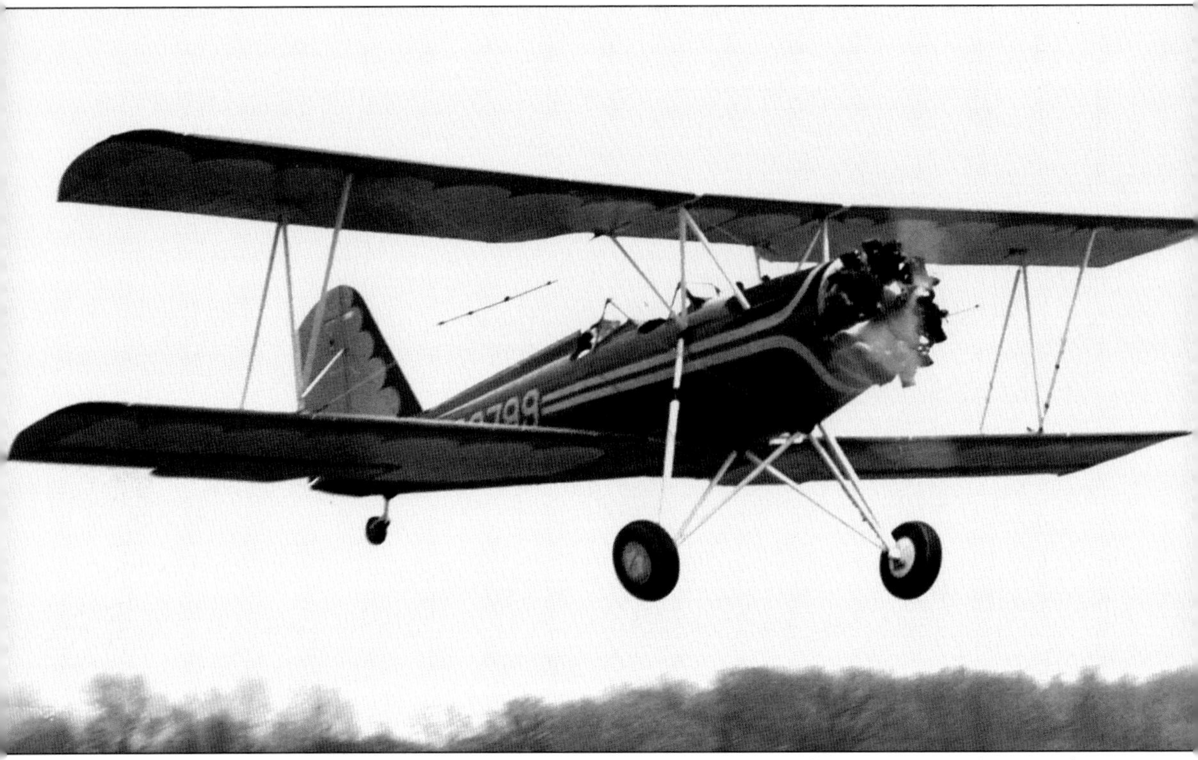

Allen Meyers, born in New Jersey in 1908, became interested in aviation after a ride in a National Guard airplane. He left home in 1928 and worked for several aircraft companies. After learning to fly in 1930, he bought an OX-5 Waco biplane. In 1931, Meyers was offered a job with Stinson Aircraft and decided to settle down giving him time to build the airplane he had been dreaming about. Built in a shop in Romulus and assembled at Wayne County Airport, it was a handsome-looking, two-place biplane with an all-metal fuselage. Meyers named it the OTW, for "Out To Win." First flight was in May 1936. In February 1941, the OTW was awarded ATC No. 736. By then, the Civilian Pilot Training Program had been activated, and there was a demand for aerobatic training aircraft. Meyers bought a farm near Tecumseh and converted it into an airport with a hangar and adjacent shop. Production began in 1941, and 102 Meyers OTWs were built. One low-wing trainer was also built, the Me-165, but it never went into production. (Photograph by the author.)

Six

World War II 1939–1945

By 1939, America had finally emerged from the Great Depression, and aircraft manufacturers were slowly getting back into production. New designs like the Douglas DC-3 had revitalized the airline industry, and airliners were now crisscrossing the continent reliably, safely, and on schedule. Flights to overseas destinations were also available in elegant-looking flying boats. Plus new training aircraft like the Piper Cub, which sold for $995 in 1940, had created renewed interest in private flying and learning to fly.

But the threat of war loomed on the world's horizons. Italy had invaded Ethiopia, and the Spanish Civil War had escalated with Germany and Russia entering the conflict. On the other side of the world, Japanese armed forces had invaded Manchuria. America remained neutral but to circumvent the neutrality policy, the Lend-Lease Act was passed in March 1941. The act allowed America to send large quantities of war goods, including aircraft, to its allies without becoming directly involved in their war.

In 1938, Pres. Franklin D. Roosevelt signed into law the Civilian Pilot Training Program with a goal of annually training 20,000 college students to fly. That program had a major impact on the light aircraft industry as flying schools began ordering new airplanes.

On December 7, 1941, Japan attacked Pearl Harbor, and America was suddenly drawn into a global conflict. Factories across America rapidly converted to wartime production, and Detroit became known as the "Arsenal of Democracy." Automobile manufacturers stopped building cars and began building tanks, guns, and other weapons of war. Between Pearl Harbor and the end of the war, America built over 300,000 military aircraft, and the miracle of Ford's Willow Run bomber plant will forever remain one of the greatest accomplishment in the history of aircraft manufacturing.

After the war, returning members of the armed services were looking forward to a bright new future. Articles in contemporary magazines predicted an exciting new world with an airplane or helicopter in every garage, and the American aircraft industry began preparing for the expected boom in sales.

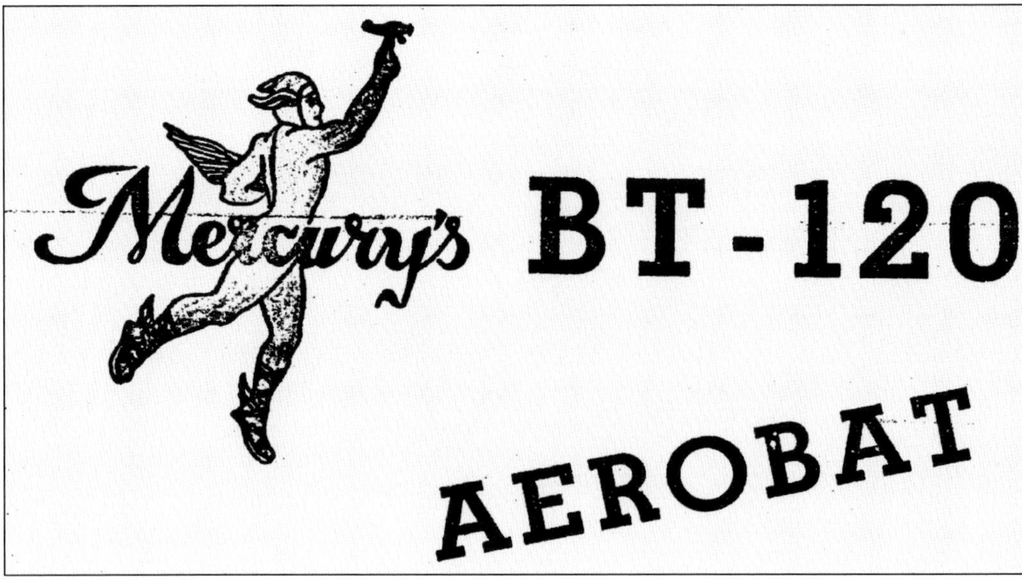

This logo for the Mercury Aircraft Corporation of Menominee, Michigan, featured the Roman god Mercury. Jack Bauman was optimistic about establishing an aircraft company in Menominee, but it failed. He left Menominee in 1941 to become chief engineer of Frankfort Sailplane Company in Illinois and later went to work for Lockheed in California. In 1945, he formed Bauman Aircraft Company and built two Bauman Brigadiers, twin-engine, six-passenger business airplanes. He retired from Lockheed in 1983.

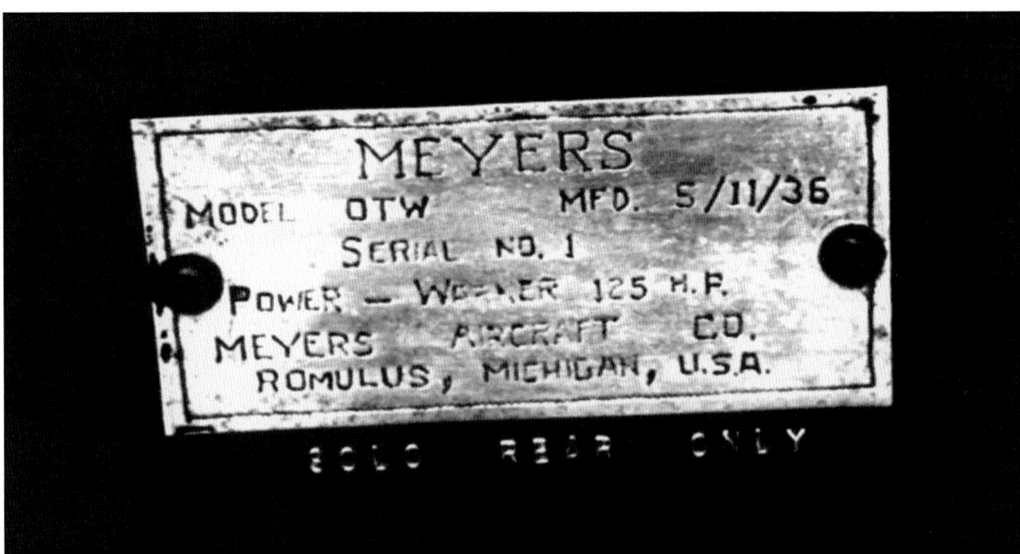

This aluminum aircraft identification plate was riveted to the instrument panel of Al Meyers's first OTW biplane, serial No. 1. The identification plate shows that the airplane was manufactured on May 11, 1936, in Romulus. Today's Federal Aviation Regulations (FAR) require that the plate be mounted externally near the tail of the aircraft and that it be made of steel. (Photograph by Everett Payette.)

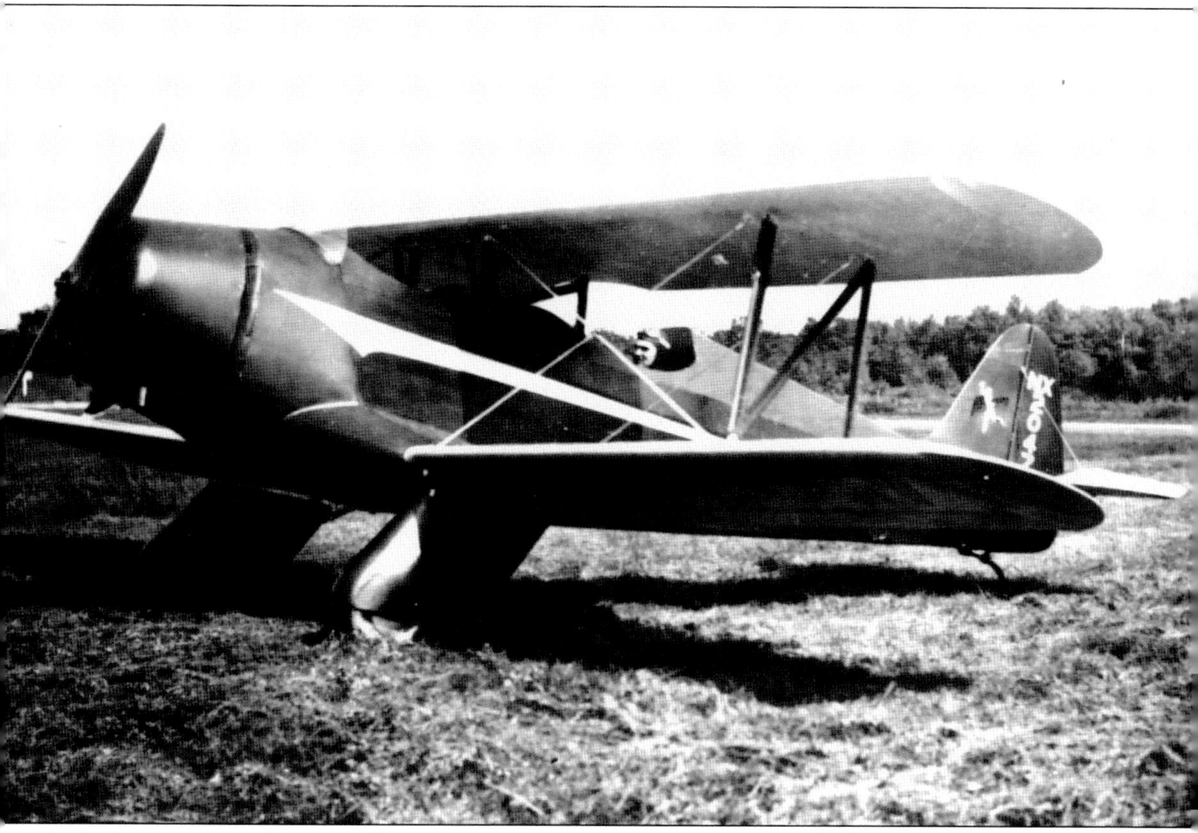

Jack Bauman was born in Knoxville, Tennessee, in 1914 and studied engineering at the University of Tennessee. In 1937, he built an original design airplane, the B-100. The four-place cabin biplane had negative stagger wings, a configuration in which the lower wing is located forward of the upper wing. After forming Mercury Aircraft Corporation, Bauman was contacted by officials from Menominee, Michigan, seeking to attract new business to their community. Bauman agreed to move there and proposed building a trainer version of his B-100 to sell to the Civilian Pilot Training Program (CPTP) flying schools. He built a new fuselage to which he attached the original B-100 wings and tail surfaces. The new Mercury BT-120 Aerobat was completed in September 1940. A second airplane was built a month later. But by then, the CPTP schools no longer needed aerobatic trainers, so Mercury Aircraft Corporation closed. A third Aerobat was built after the war by two local pilots using left over parts. Bauman died in February 1996 after a long career in aviation.

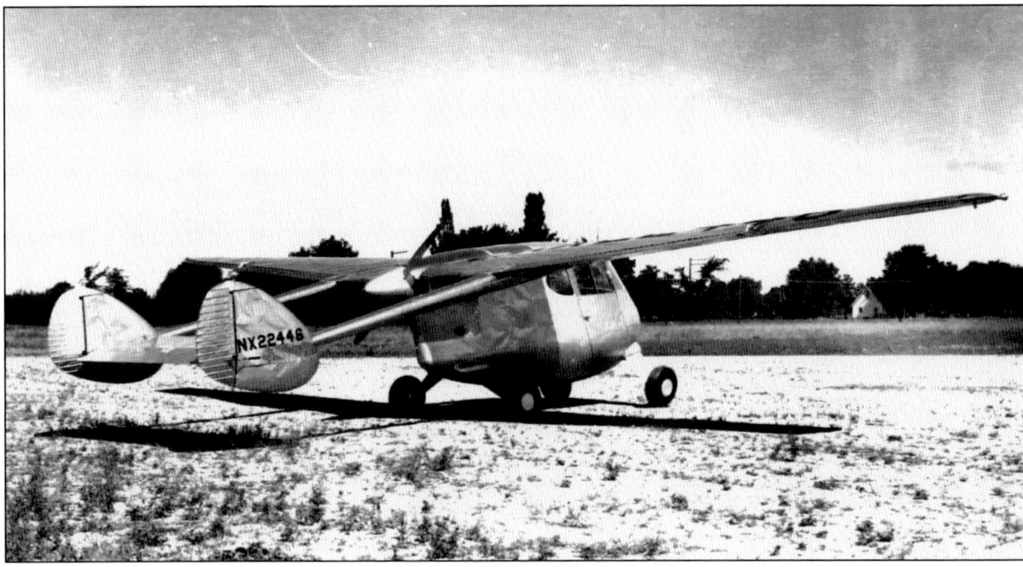

In 1942, Bill Stout built this Skycar II, developed from his earlier Skycar of 1931. It was constructed of stainless steel and first flown in late 1943 at Mettetal Airport in Plymouth, Michigan. Stout received a U.S. Army Air Corps contract for the Skycar II, designated XC-65, which was to serve as the prototype for a large military transport, the XC-107. After the XC-65 was destroyed in a hangar fire, both contracts were canceled.

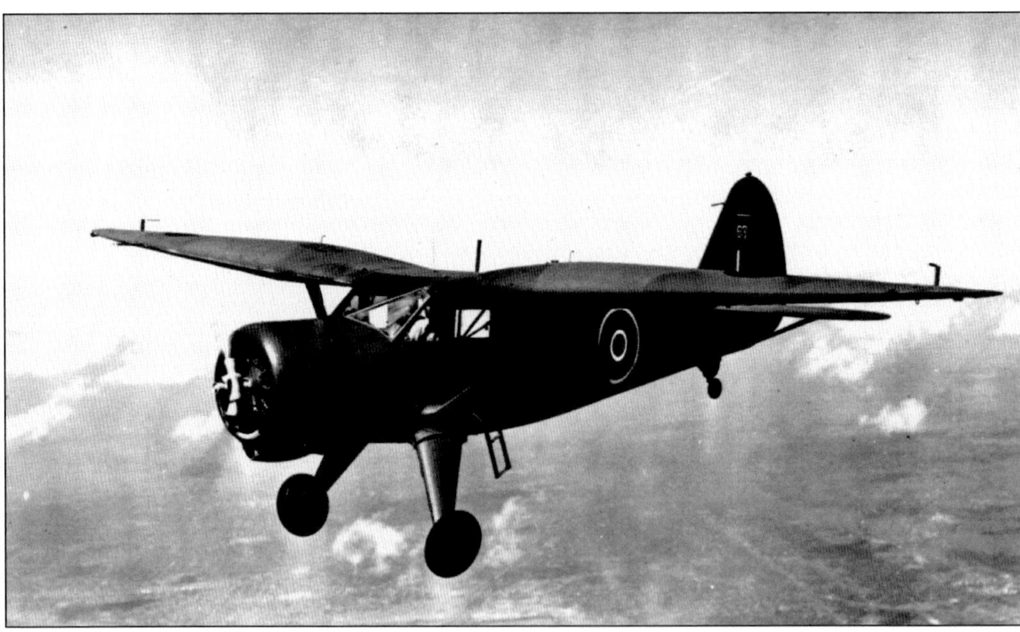

After Eddie Stinson's death in 1932, Cord's Lucius Manning was named president of Stinson Aircraft. In 1934, Aviation Corporation (Avco) bought Cord and Stinson, and Stinson Aircraft became a division of Avco. In October 1939, Avco merged Stinson Aircraft and Vultee Aircraft and Stinson moved some of its operations to a factory in Nashville. However, when England ordered 500 SR-10J Reliants for navigation training in 1942, all 500 were manufactured in the Wayne, Michigan, factory.

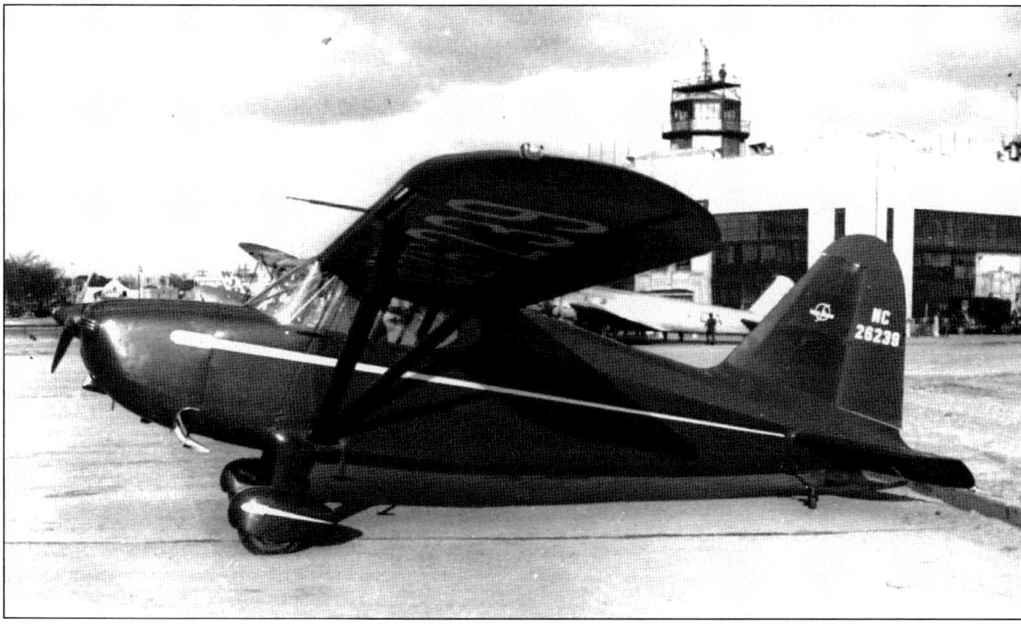

Stinson entered the lightplane market in 1939 with the three-place model 105 powered by a 75-horsepower Continental engine. Priced at $2,995, it was an instant success (seen above). Production began in Nashville but when Vultee Aircraft needed space to build military aircraft, Stinson Aircraft moved back to Michigan. By the time production ended in 1940, Stinson Aircraft had built 1,035. In 1940, the U.S. Army requested bids for a small observation airplane, so engineer A. P. Fontaine modified a model 10A Voyager, converting it to tandem seating and installing a 100-horsepower Lycoming engine. A second version with a 175-horsepower engine won the competition and became the prototype for the L-5 Sentinel. The production L-5 was widely used during World War II and the Korean Conflict. Stinson Aircraft built approximately 3,285 L-5s between 1942 and 1944. (Below, courtesy Peter M. Bowers.)

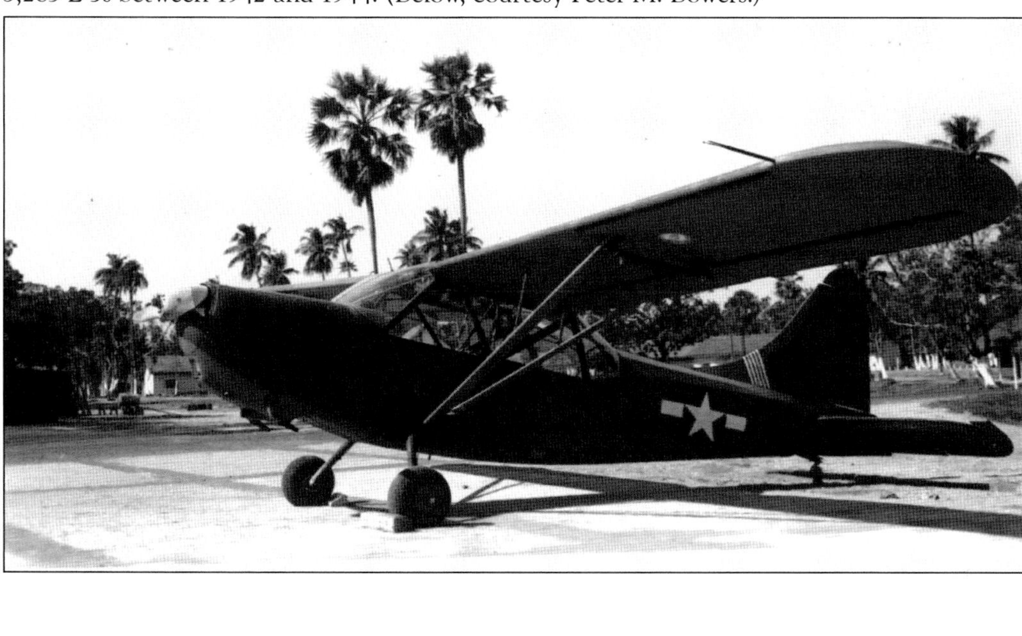

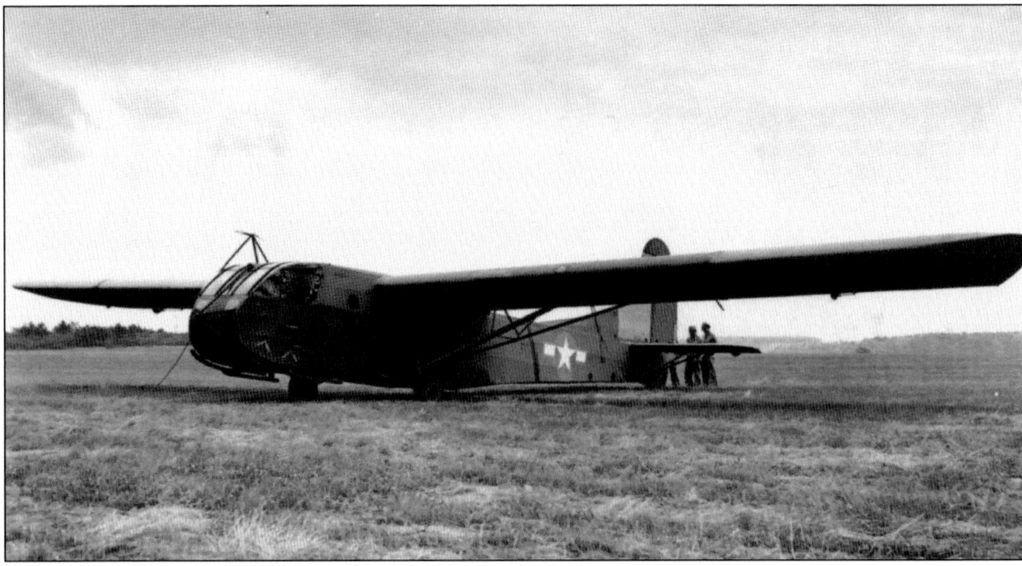

The German army introduced a new form of warfare in 1940 when glider troops captured a strategic fort in Belgium. The U.S. Army had no troop gliders in 1940, so a program was initiated to design a troop glider, organize a glider corps, and train glider pilots. The Waco Aircraft Company of Ohio won the design competition with its 15-seat CG-4A. Some 16 manufacturers were involved in the production program including Ford Motor Company. Ford built 4,190 gliders (seen above) in a Ford factory in Iron Mountain, Michigan. One other Michigan manufacturer, the Gibson Refrigerator Company of Greenville, built 1,078 CG-4As. A Gibson-built troop glider was the first to land in France on D-day. Waco later designed a larger, 42-seat troop glider, the CG-13A. Ford was awarded the contract and built one prototype in Dearborn (seen below) and 97 production models in Iron Mountain.

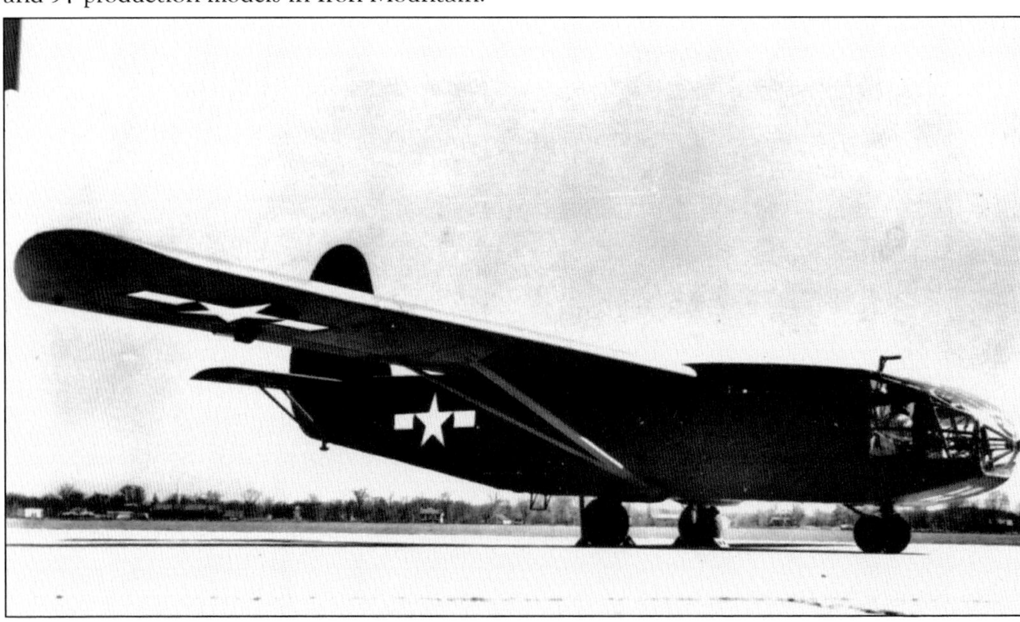

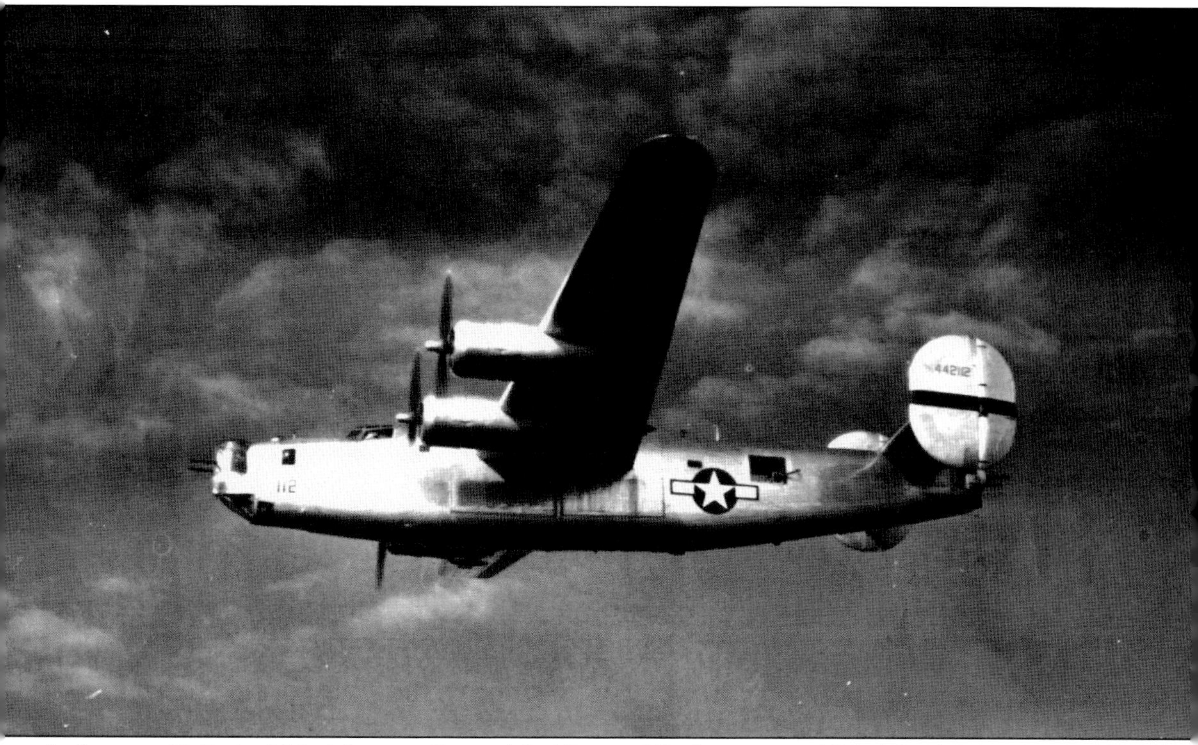

In January 1941, the U.S. Advisory Council for National Defense invited Edsel Ford to a meeting with Consolidated Aircraft Corporation officials. The government wanted Ford to become involved in aircraft production by building components for Consolidated's B-24 Liberator bombers. But Ford proposed building complete airplanes on an automobile-type production line. In April 1941, a site at Willow Run, Michigan, was selected for a new factory, and by the end of that year, a 2.5-million-square-foot building had been erected and a mile-square airport was built. On April 15, 1942, the first B-24 rolled off the end of the mile-long assembly line. Production increased rapidly from that day on and reached a peak in April 1944. That month, with 25 working days and two nine-hour shifts, Ford built 455 B-24 bombers—or one bomber every 59.3 minutes. Production ended in June 1945, after Ford had built 6,792 bombers. Some sources show 8,685 built, but that total includes 1,893 "Knock Down" kits fabricated by Ford but assembled at factories in Tulsa and Ft. Worth. (Photograph by the author.)

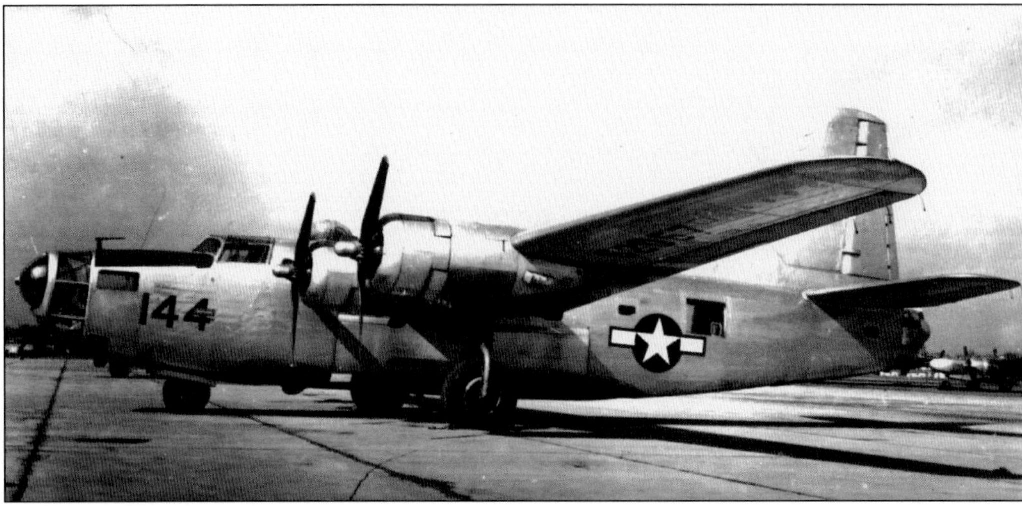

A total of 18,482 B-24 Liberator bombers were produced in five different factories, including Willow Run. The four-engine B-24 had a gross weight of 65,000 pounds and a wingspan of 110 feet. All production B-24s had the distinctive twin-tails except for seven YB-24Ns built by Ford in June 1945 with single vertical tails (seen here). If the war had continued, all future B-24s would have been single tail models.

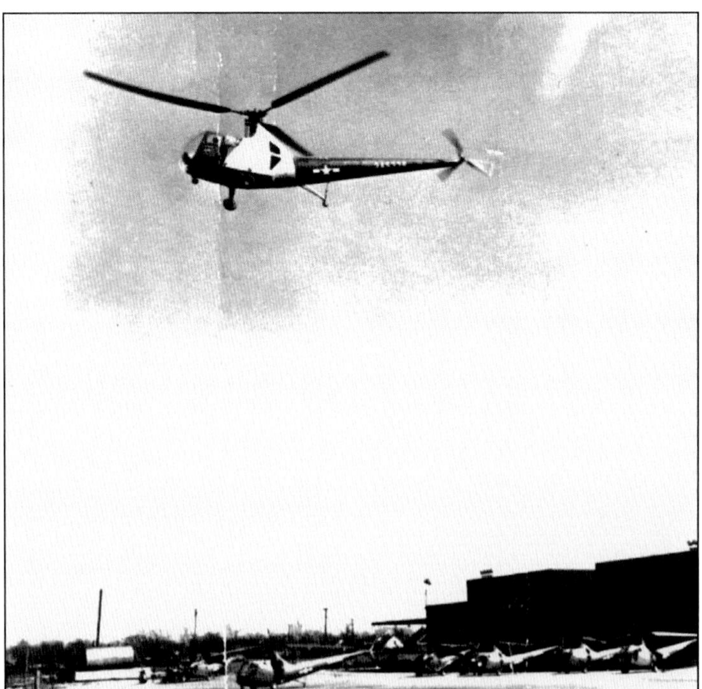

Igor Sikorsky, a Russian immigrant, arrived in America in 1918 and in 1923 founded Sikorsky Aircraft, a leading manufacturer of flying boats. In 1939, Sikorsky built the world's first successful helicopter, the VS-300. A military version, the R-4, was built in large quantities. In 1944, a contract for an improved R-4, the R-6 Hoverfly, was awarded to Nash-Kelvinator in Detroit. They assembled 219 from components fabricated in a factory in Grand Rapids.

Seven

POSTWAR BOOM AND BUST 1946–1954

In 1945, a highly respected organization conducted a public opinion poll and reported that 32 percent of the American adult population wanted to own an airplane. In addition, seven percent stated that they definitely planned to buy one after the war had ended. Based on those predictions aircraft manufacturers began preparing for a postwar sales boom of close to three million airplanes.

By early 1946, General Aviation manufacturers like Aeronca, Beech, Piper, and Stinson had all introduced new models aimed at winning a share of that market. In addition, major airframe manufacturers like North American and Republic had entered the private aircraft market as well as a number of new manufacturers with fresh, new designs.

But the pollsters had been wrong. Returning servicemen were interested in raising a family, buying a home, or attending college under the GI Bill, not in buying airplanes. However, in 1946, sales were booming, giving credence to the pollster's predictions with sales reaching 34,870 units. But by 1947, sales had dropped to 15,600 and by 1948 to 7,040. By 1951, only 2,300 private airplanes had been sold. The bust has been attributed to reliance on questionable market predictions plus overproduction, but other factors were involved. The Berlin Airlift and the Korean War had Americans concerned about the future and escalating prices for new airplanes had put ownership beyond reach of the average wage earner.

Nevertheless Michigan's aircraft manufacturers fared well during this period. Stinson was selling a record number of airplanes, the Willow Run bomber plant was again building airplanes for the U.S. Air Force, and Chrysler's Missile Division was building Redstone and Jupiter missiles for the U.S. Army. Furthermore, William Lear of Grand Rapids had been awarded the Collier Trophy for his Autopilot Coupler System, and Kimber Sigler, Michigan's "Flying Governor," promoted aviation by piloting his own airplane on state business.

Meanwhile Paul Poberezny had formed the Experimental Aircraft Association (EAA) for individuals who were designing and building their own airplanes The organization grew rapidly, as new designs were introduced that could be built by the average person with minimum skills. A new era of aircraft manufacturing was about to begin.

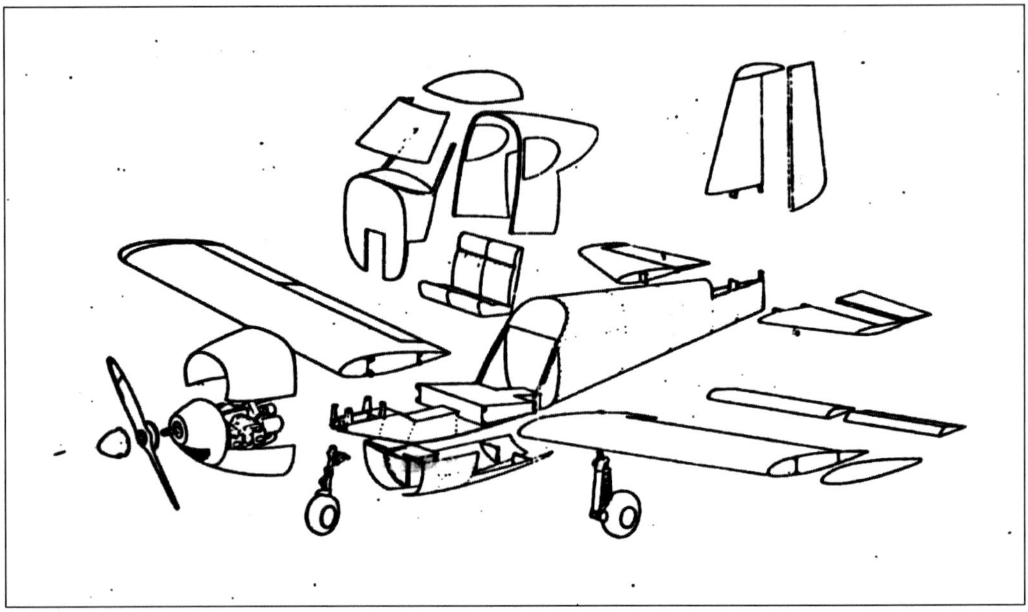

Bendix Corporation planned to build their new model 55 on an automobile-type assembly line in Garland, Texas. The airplane's major components were designed for simplicity and low cost, as seen in this Benoix drawing, P-52-107. Each wing panel had only 19 parts, not counting the skin. The rudder and both elevators were identical, as were the fin and horizontal stabilizers. In 1946, 100 designers were working on the production tooling when the program was canceled.

Seen on top is the original Enstrom Corporation's logo. In 1971, the company, located in Menominee, Michigan, changed its name to include the word "helicopter" as seen in the logo on the bottom. Enstrom has been building helicopters for almost 50 years, longer than any other Michigan aircraft manufacturer. The company builds the piston-powered F28F and 280FX and the turbine-powered 480B, and as of 2008 has built a total of 1,130 helicopters.

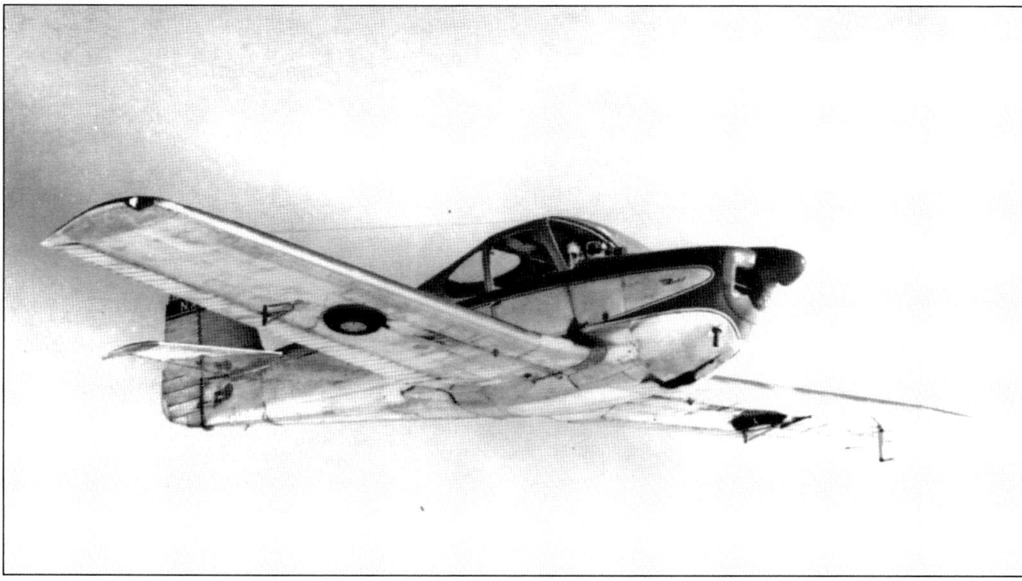

In 1944, Ernest Breech, president of Bendix Corporation, formed an aircraft division and planned to enter the postwar lightplane market. Breech recruited several key Stinson Aircraft engineers, and they designed the model 55, a two-seat, all-metal airplane (seen above). It was first flown in December 1945 at Windsor Airport, across the Detroit River in Canada. Several months later, a second model 55 joined the flight test program at Willow Run Airport. Meanwhile Bendix built a four-place, twin-boom pusher, the model 51 (seen below), plus model 51A, an amphibious version of model 51. But in May 1946, Breech left Bendix Corporation to become vice president of Ford Motor Company. The new Bendix management was not enthusiastic about building airplanes and closed the aircraft division. The four airplanes were donated to local schools and eventually scrapped.

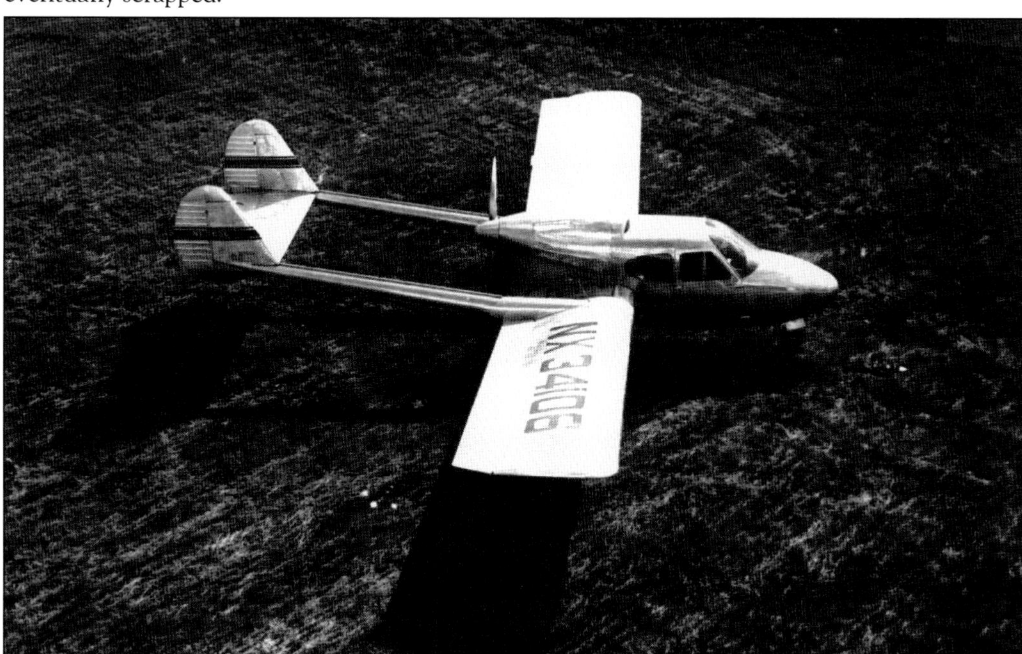

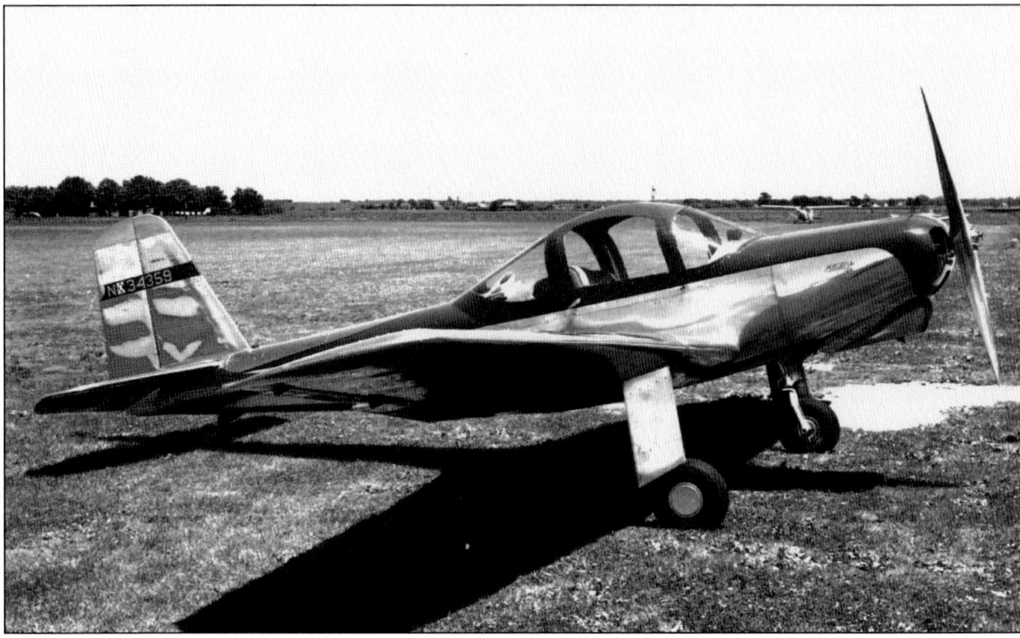

When production of OTW biplanes ended, Allen Meyers began planning for the postwar market. While Meyers was developing a new airplane Meyers Aircraft built 12 Dart GCs for aircraft brokers Applegate and Weyant. The two-place Meyers MAC-125 prototype was test flown by Meyers, but when he could not recover from a spin, he bailed out. In 1948, a revised version, the MAC-145 (seen here), flew successfully, and 20 were manufactured.

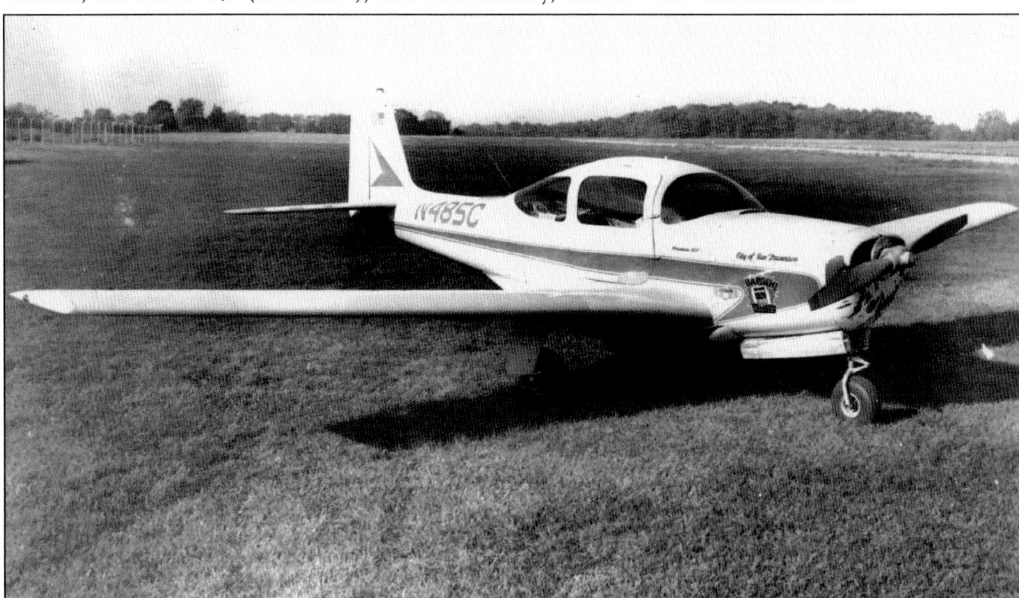

In 1958, Meyers Aircraft introduced the MAC-200, a four-place, all-metal, airplane with a cruising speed of over 200 miles per hour. One MAC-200 was built with a "wet wing" and set a solo, around-the-world record in 1959 (seen here). Other MAC-200s won numerous stock-class air racing events. After building 45 MAC-200s, Meyers sold the design rights to North American Rockwell in 1965. Meyers suffered a stroke in 1971 and died in 1976.

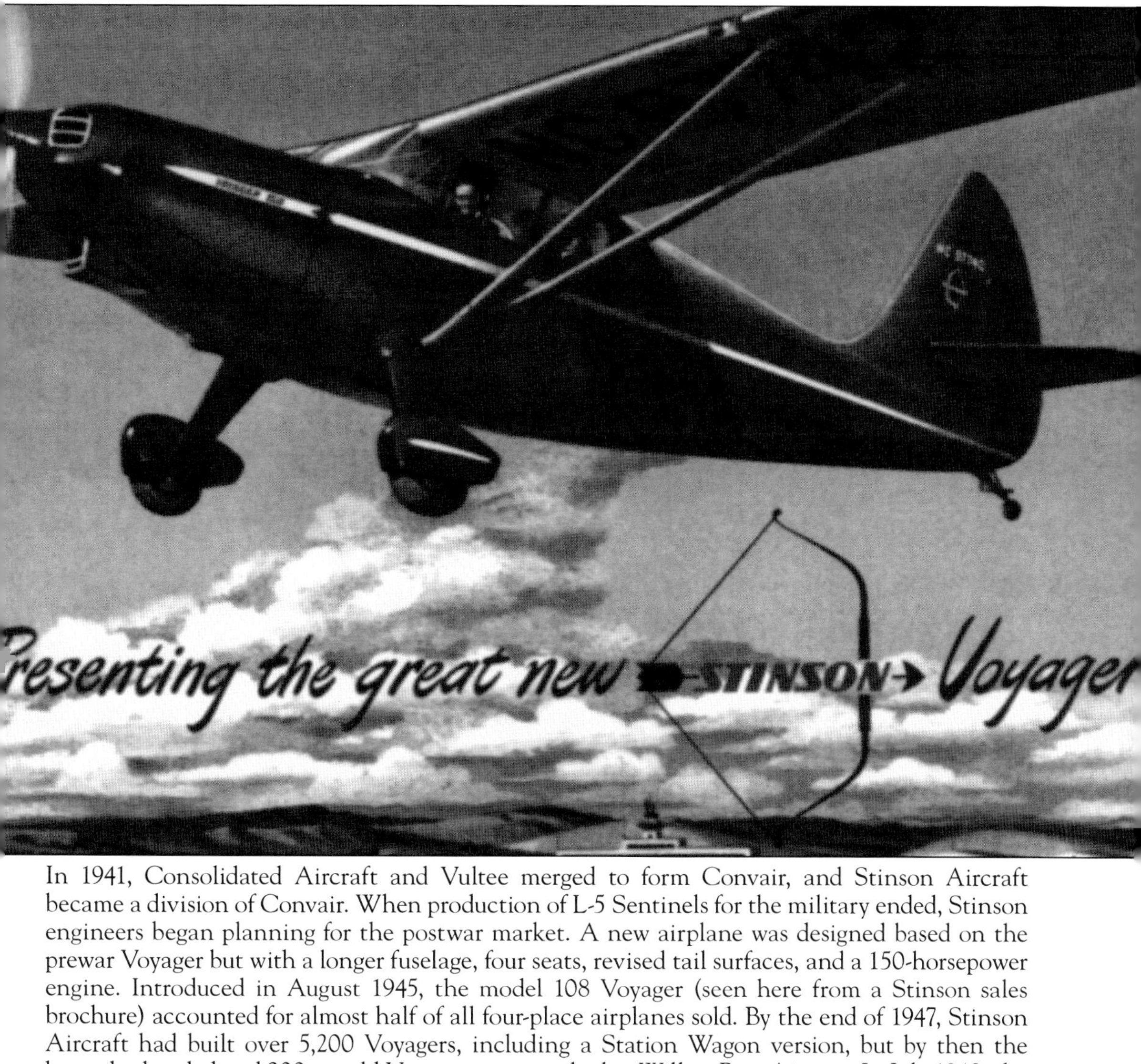

In 1941, Consolidated Aircraft and Vultee merged to form Convair, and Stinson Aircraft became a division of Convair. When production of L-5 Sentinels for the military ended, Stinson engineers began planning for the postwar market. A new airplane was designed based on the prewar Voyager but with a longer fuselage, four seats, revised tail surfaces, and a 150-horsepower engine. Introduced in August 1945, the model 108 Voyager (seen here from a Stinson sales brochure) accounted for almost half of all four-place airplanes sold. By the end of 1947, Stinson Aircraft had built over 5,200 Voyagers, including a Station Wagon version, but by then the boom had ended and 200 unsold Voyagers were parked at Willow Run Airport. In July 1948, the factory was closed and five months later Convair sold Stinson Aircraft to Piper Aircraft. Earlier, in 1946, Stinson Aircraft had built one experimental, all-metal, four-place airplane, the model 106 *Skycoach*, but it never went into production. The company that Eddie Stinson had founded 22 years earlier had built over 12,000 airplanes, but was now only a memory.

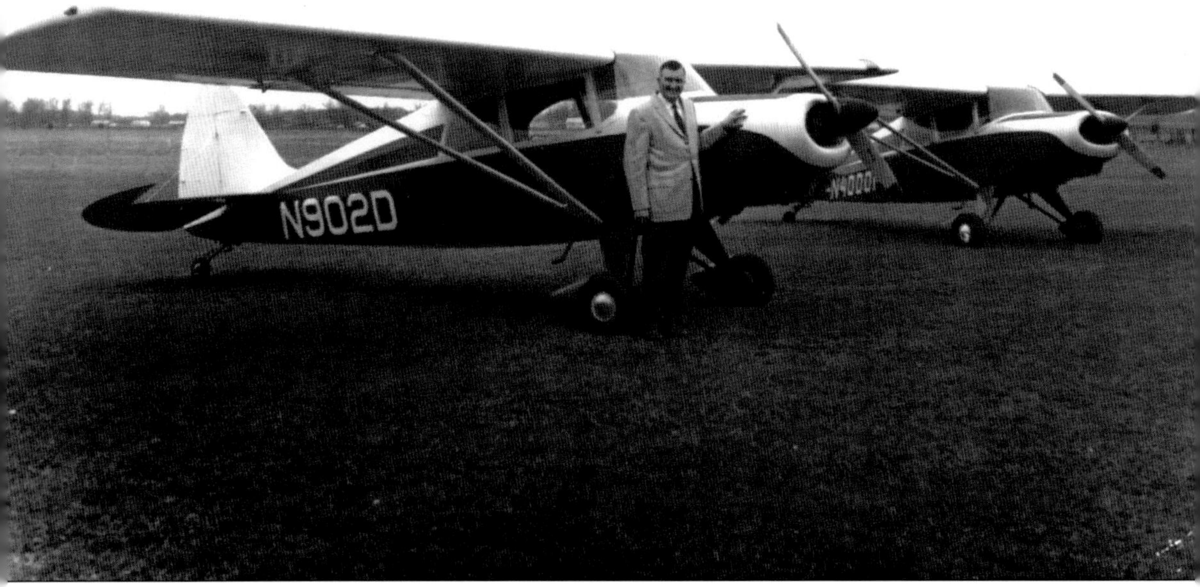

Belford D. (B. D.) Maule was born on a farm in Ohio in 1911. In 1930, B. D. joined the U.S. Army Air Corps as an aircraft engine mechanic. After leaving the U.S. Army Air Corps, he built a single-place airplane and taught himself to fly. In 1940, he moved to Jackson, Michigan, where he manufactured aircraft accessories. At Jackson, B. D. built two ornithopters, flapping-wing aircraft, and experimented with the concept. In 1947, he bought a farm near Napoleon, Michigan, and converted it into an airport. There, B. D. built a four-place, cabin monoplane, the BD-4, which he test flew in 1957. A 210-horsepower engine gave it "Short Take Off and Landing" (STOL) performance that appealed to grass roots pilots. In 1961, B. D. and wife June formed the Maule Airplane Company to manufacture the airplane. Soon a flood of orders was arriving from pilots in this country and from overseas. By 1960, Maule had 60 employees building two BD-4 Jetasins a week in a crowded factory. When Moultrie, Georgia, officials invited B. D. to move the company from Michigan to a large hangar located on Spence Field, he accepted their offer. After building 249 BD-4s, Maule Aircraft moved to its present location in Georgia. B. D. died in 1995.

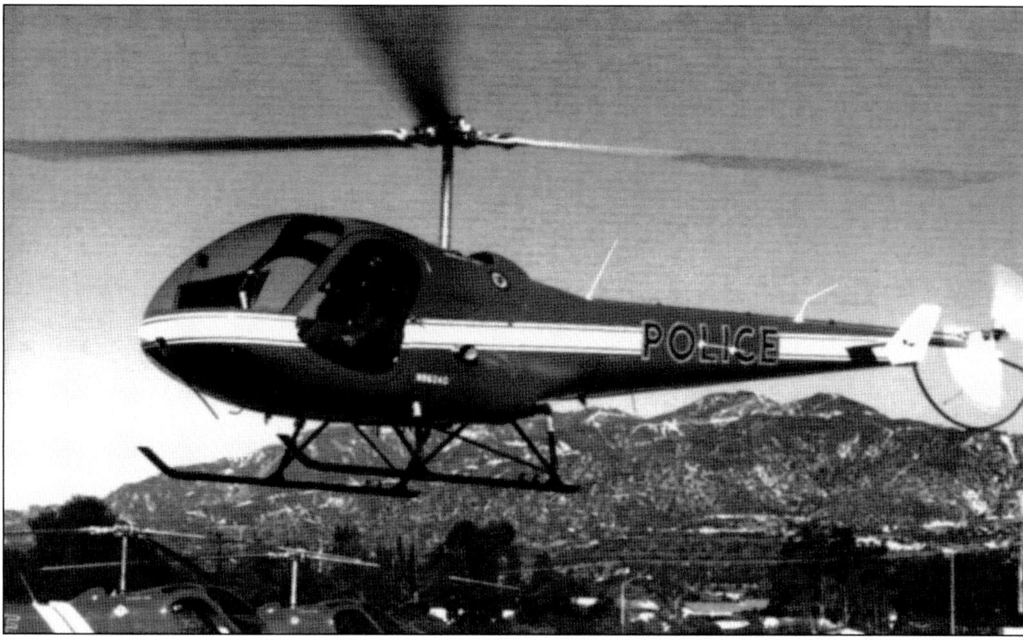

In the 1950s, master mechanic Rudy Enstrom built an original design helicopter in his home north of Menominee, Michigan. It flew successfully so Jack Christensen, a Menominee businessman, sold stock and formed a company to build Rudy's helicopter. Since then Enstrom has been owned by the Purex Corporation, Attorney F. Lee Bailey and Bravo, a Dutch company. Now privately owned, Enstrom has built 1,130 helicopters that are operating in 45 countries around the world.

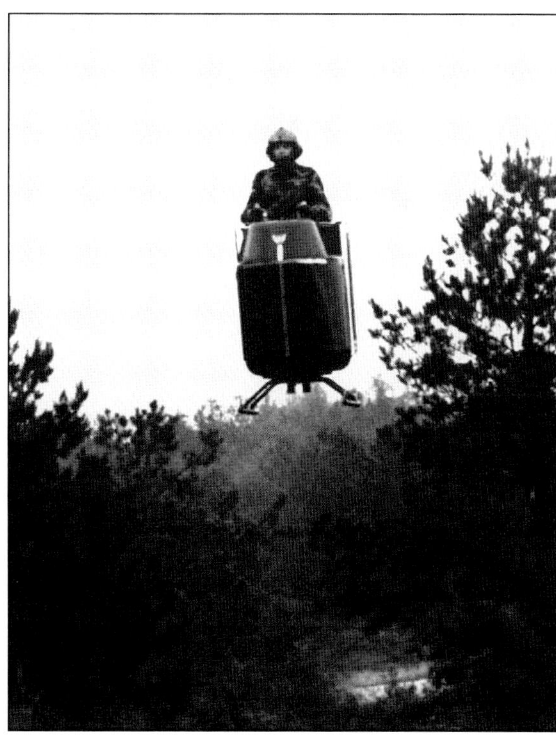

Dr. Sam Williams founded Williams International in 1954. The Walled Lake, Michigan, company builds turbofan engines for cruise missiles and for business jets. Dr. Williams was awarded the Collier Trophy in 1978 for his pioneering work in aircraft propulsion. The company has built a number of Williams Aerial Systems Platforms (WASPs) that were evaluated by the U.S. Army in the 1980s. They can fly at 60 miles per hour and have a flight duration of 45 minutes.

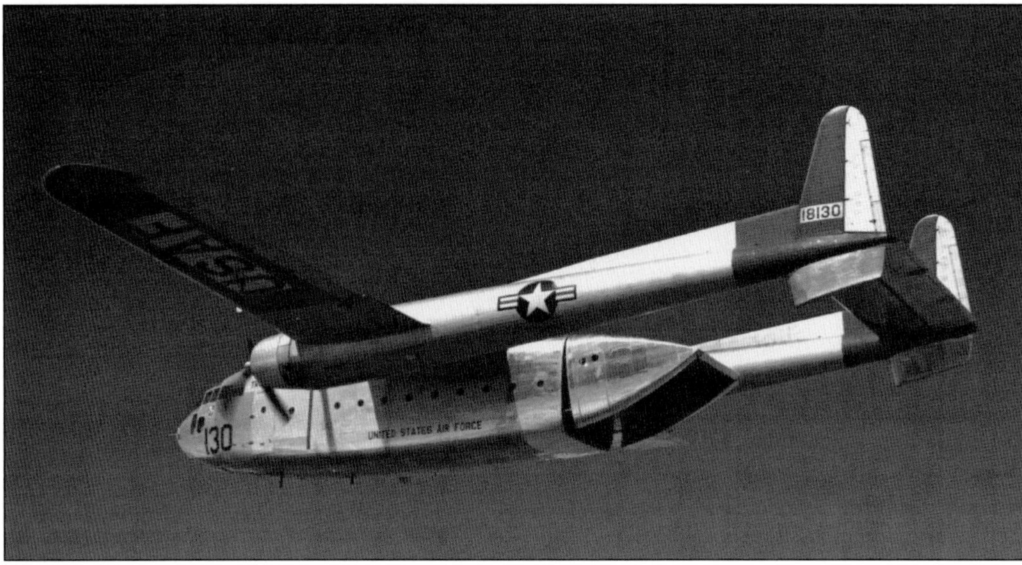

After the start of the Korean War, the U.S. Air Force awarded a contract to Kaiser Industries to build the Fairchild-designed C-119F Flying Boxcar in the Willow Run bomber plant, then owned by Kaiser. Production began in 1952, and 71 were built (seen here). In 1952, Kaiser Industries bought Chase Aircraft including a contract to build 300 Chase C-123B Providers. Kaiser Industries built five at Willow Run, but the U.S. Air Force canceled the contract because of poor workmanship. (Courtesy Dave Ostrowski collection.)

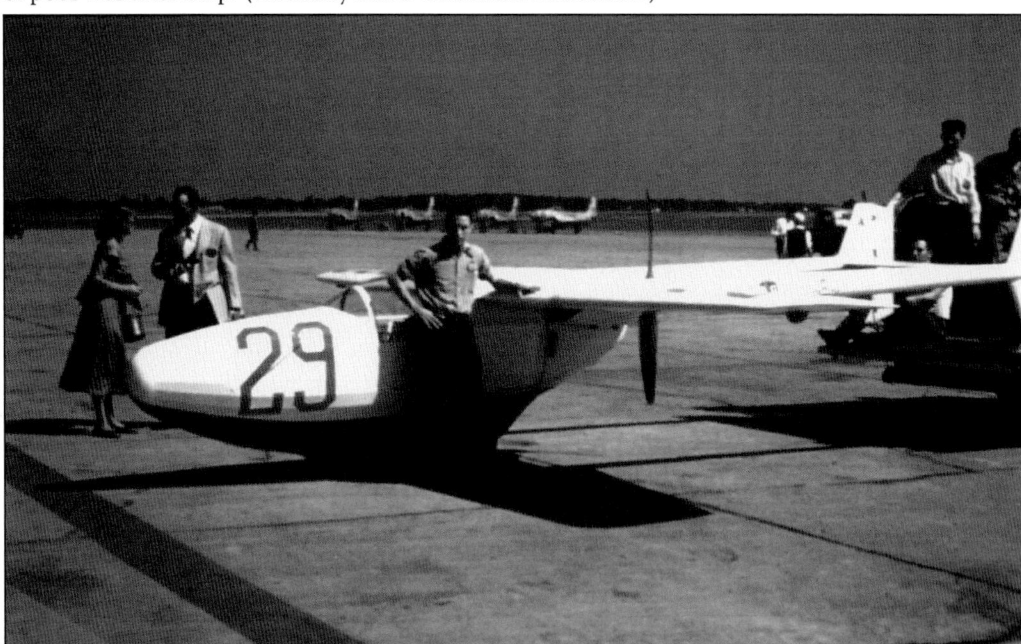

Aeronautical Engineering students at Lawrence Institute of Technology (LIT), Highland Park, Michigan, designed and built this Goodyear-class raceplane. The project was supervised by Prof. George Martin, who had built gliders as an LIT student before the war. The LIT racer qualified at 127.9 miles per hour at the 1949 Cleveland Air Races, piloted by Carlton Ambler (seen here). The airplane is presently on display at the school's Southfield, Michigan, campus.

Eight

The Homebuilt Movement 1955–2009

In 1953, Paul Poberezny, an ex–U.S. Air Force and Wisconsin Air Guard pilot, had formed the Experimental Aircraft Association (EAA) to coordinate activity between designers and builders of experimental private airplanes. The organization grew rapidly and soon attracted sport and recreational pilots and eventually expanded to encompass all segments of general aviation from ultralights to warbirds.

As the EAA grew, talented new members began introducing innovative designs of all-metal and later composite construction. Men like John Thorp, Jim Bede, Burt Rutan, and Dick VanGrunsven were creating exciting new airplanes that were sleeker-looking, faster, and more fun to fly than the factory-built airplanes then on the market.

The EAA, working in close cooperation with the Federal Aviation Administration (FAA) also promoted regulations that were of interest to private and recreational pilots. The Sport Pilot Certificate and the development of Light Sport Aircraft (LSA) made it easier for enthusiasts to earn a pilot's license. In addition, the EAA promoted the use of automobile fuels as a cost-saving measure. Another major EAA accomplishment is the Young Eagles program that has introduced over 1.5 million youngsters to the world of flight.

But EAA has not forgotten its roots—the experimental airplane builder. At AirVenture 2008, the EAA celebrated the completion of the 30,000th homebuilt airplane, an RV-9 built by Bob Null. Furthermore, one kit manufacturer, Van's Aircraft in Aurora, Oregon, recently announced the completion of the 6,000th airplane built from a Van's Aircraft kit. According to EAA records, between 1,000 and 1,200 amateur-built airplanes are licensed by the FAA each year, and the numbers continue to increase.

There are approximately 1,400 homebuilt aircraft registered in Michigan, but if one counts back to 1953 when EAA was founded, the total number of homebuilts would be closer to 2,000. With such a large number of airplanes ranging from standard kit-built types to sailplanes, helicopters, aerobatic aircraft, and one-only experimental types, it would be impossible to depict them all in the pages of this book. As a result, this chapter shows only a sampling of the many homebuilt airplanes built in Michigan.

The original logo for the EAA (seen at left) reflected the organization's early emphasis on experimental aircraft. As the EAA expanded to include a broader spectrum of general aviation, the logo was changed to the present design (seen below). Today the EAA has 170,000 members in 105 countries and over 1,000 local chapters worldwide. EAA's AirVenture convention and fly-in, held annually at Wittman Regional Airport in Oshkosh, Wisconsin, attracted over 540,000 visitors and more than 10,000 airplanes in 2008.

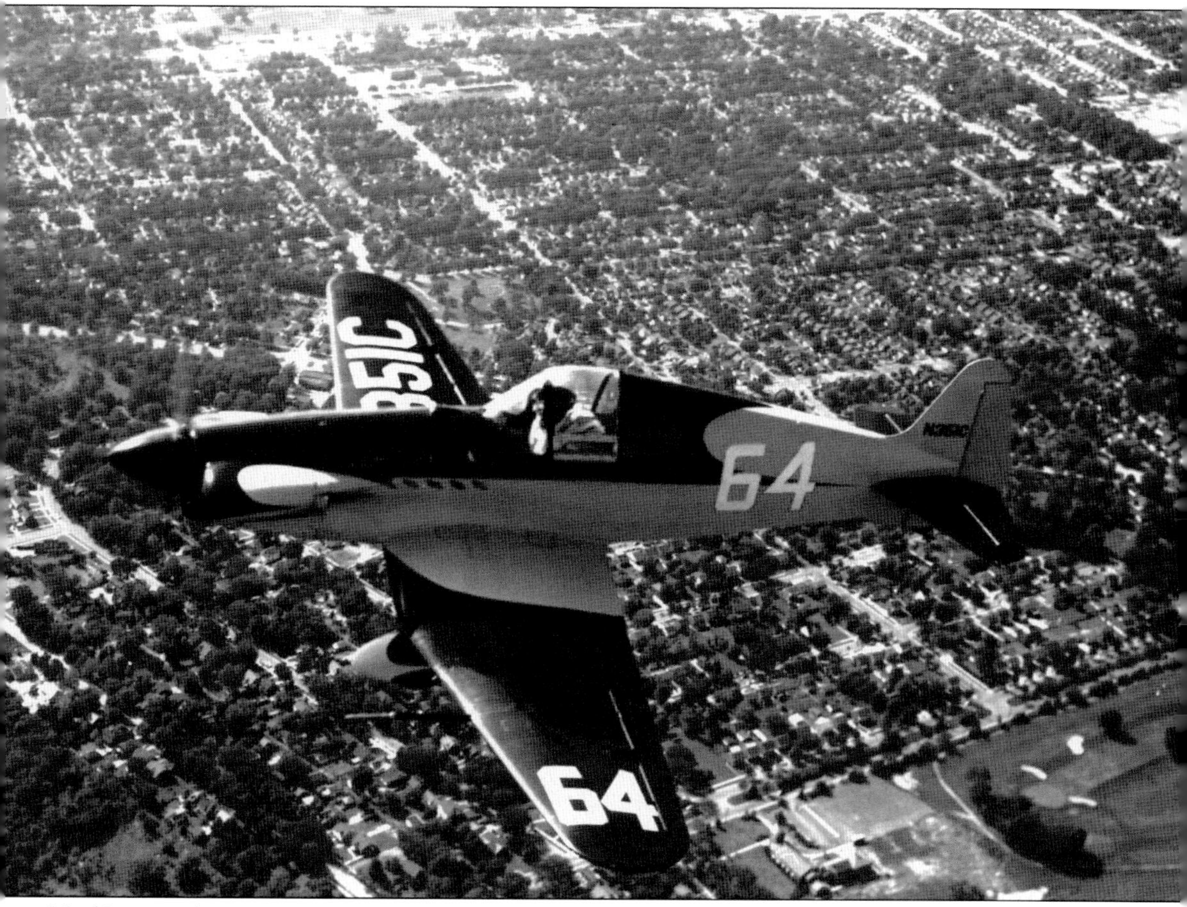

Neal Loving, born in Detroit in 1916, built model airplanes and had dreams of becoming a pilot. Friends cautioned him that African Americans were not welcome at flying schools, but Loving was determined to learn to fly and soloed in 1939. In 1941, Neal built a glider of his own design, but during a test flight, he stalled and crashed, losing both legs below the knee. By 1946, he had returned to flying using artificial legs and opened a flying school at Detroit City Airport. Loving studied engineering at Wayne State University and built an all-wood raceplane. He test flew *Loving's Love* in 1950 and participated in the 1951 Detroit Air Races. In 1956, Loving made a 1,400-mile flight to Jamaica in *Loving's Love,* where he met his wife Clare. They later moved to Dayton, where Loving designed supersonic aircraft for the U.S. Air Force. Loving died in December 1998; he was the personification of the dedicated homebuilder, having overcome major obstacles to build three airplanes of his own design. (Courtesy Dave Ostrowski collection.)

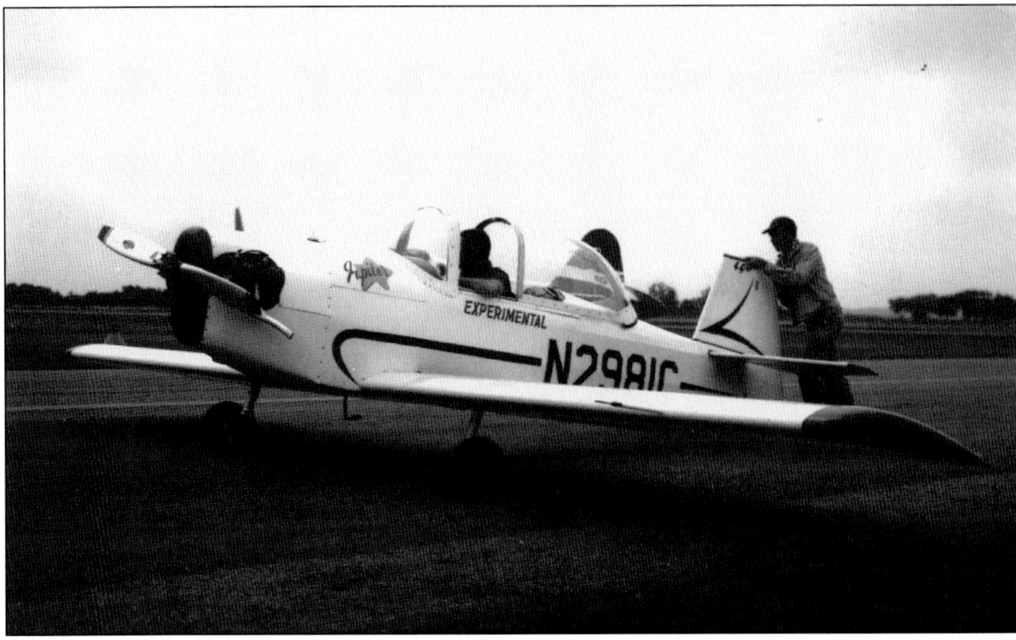

Ken Champion, a machinist from Gobels, Michigan, designed and built the Jupiter J-1 in 1959 (seen here). It had an all-wood framework covered with a fiberglass skin. A second airplane, the Jupiter K-2, was built in 1963 of stressed-skin laminated fiberglass construction. Champion was an early pioneer in adapting plastic and fiberglass materials to aircraft construction. (Photograph by Art Krieger.)

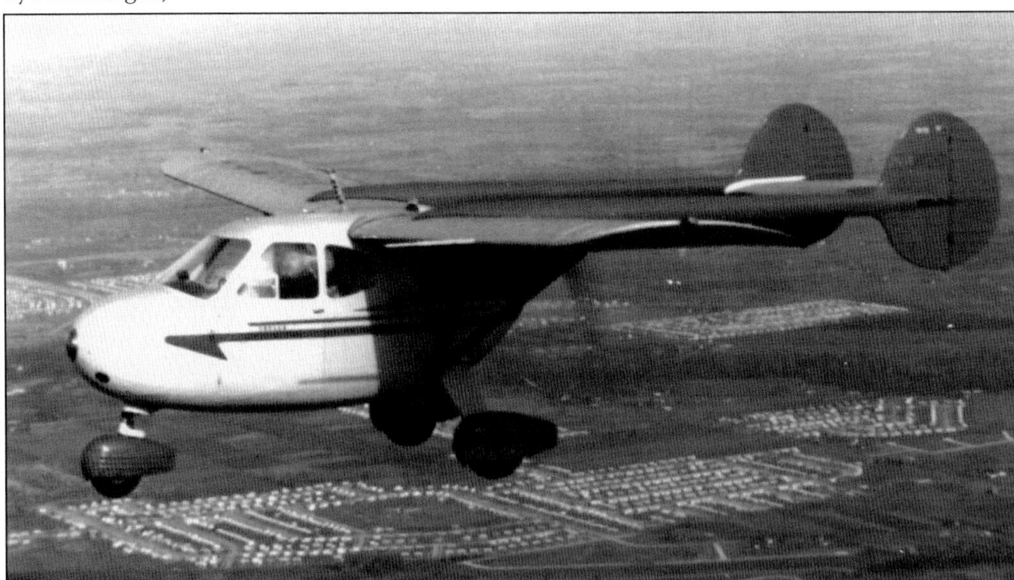

Frank Trella and his brothers displayed their Trella Speedster at the 1928 Detroit Aircraft Show but received no production orders due to the Great Depression. Trella later worked for Buhl Aircraft and Stinson Aircraft and during the war for Waco in Ohio. Trella returned to Detroit and designed the T-106, an all-metal, twin-boom "pusher." First flight was in September 1949. Only one was built, which is now in storage in Florida. (Photograph by the author.)

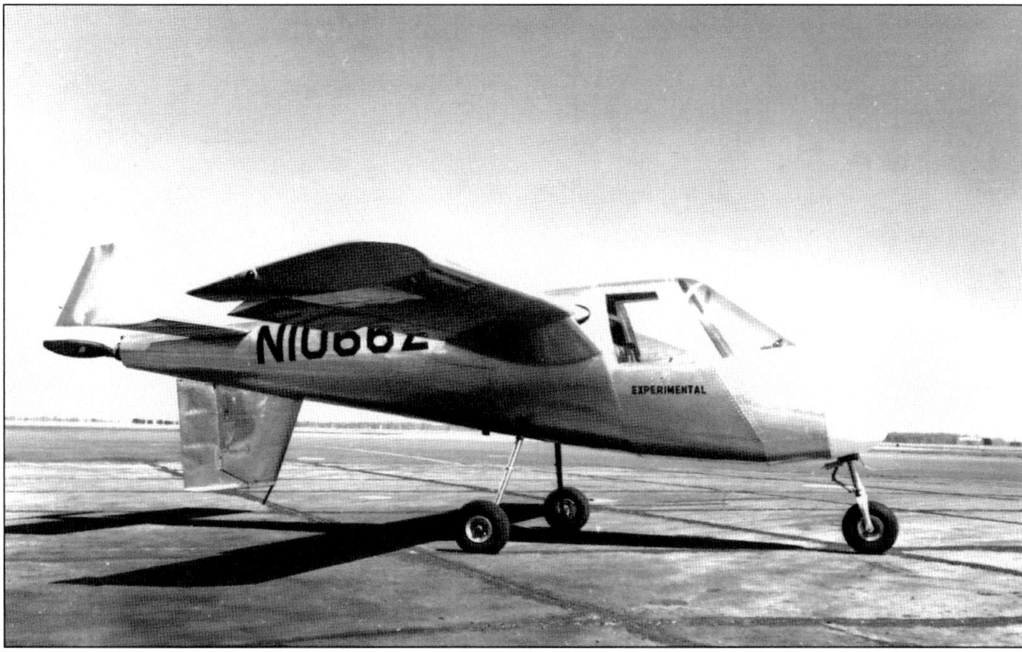

Prof. Edgar Lesher, born in Detroit in 1914, joined the University of Michigan's Aerospace Department in 1942. His aircraft design and flight test courses were widely recognized for their excellence. In 1961, he designed and built the Nomad (seen above), a two-place, all-metal airplane with a mid-mounted engine driving a pusher propeller at the rear via an extension shaft. A second airplane, the Teal, was a single-place version of the Nomad with retractable landing gear (seen below), designed to break existing Federation Aeronautique Internationale (FAI) records. Between 1967 and 1976, Lesher set speed, distance, and altitude records for aircraft weighing less that 500 kilograms (1,102 pounds), for which he was awarded the Louis Blériot Medal by the FAI. Professor Lesher died in 1998. (Above, photograph by the author; below, photograph by Philip Pauley.)

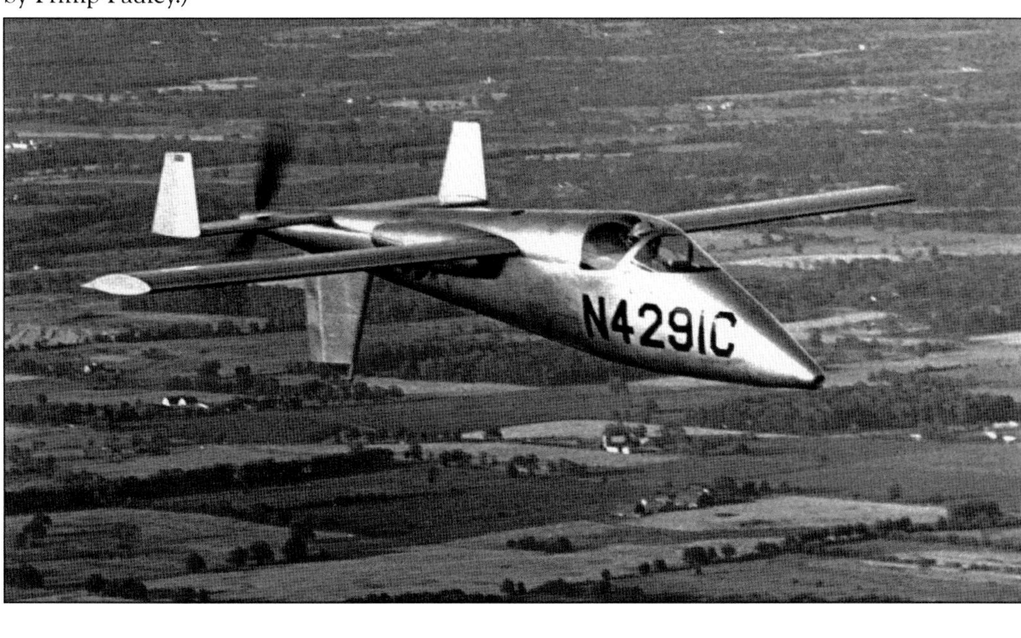

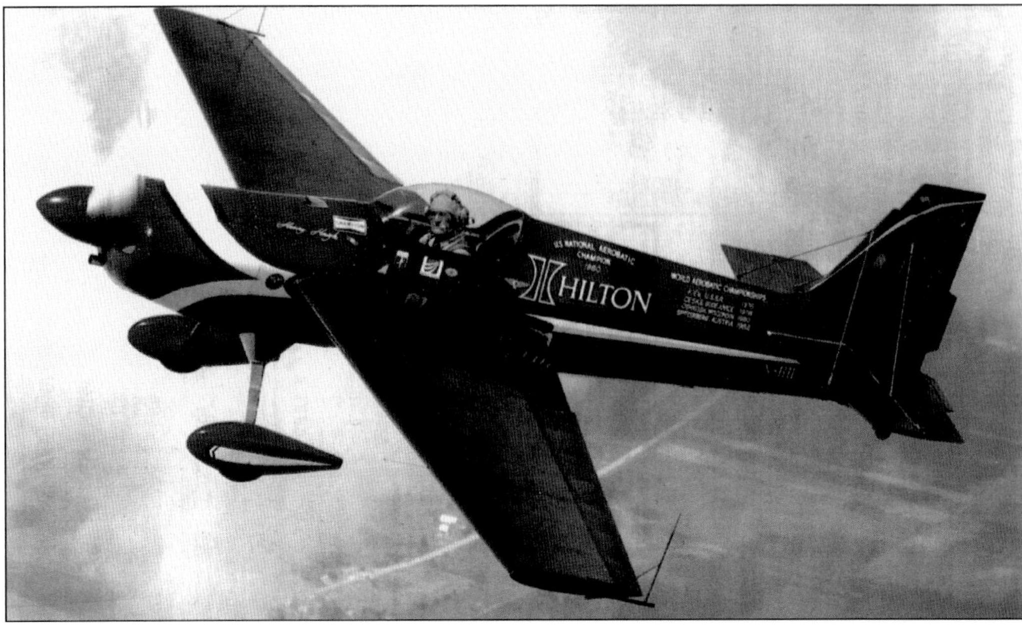

Henry Haigh, born in 1924, was a champion model airplane builder in his teens. He joined the U.S. Army Air Corps in 1942 and flew B-24 and B-29 bombers. After the war, he became active in aerobatic competition and eventually reached championship level. As a member of the United States Aerobatic Team, Haigh participated in eight world aerobatic contests. In 1988, he won the World Aerobatic Championship flying his homebuilt *Superstar* (seen here).

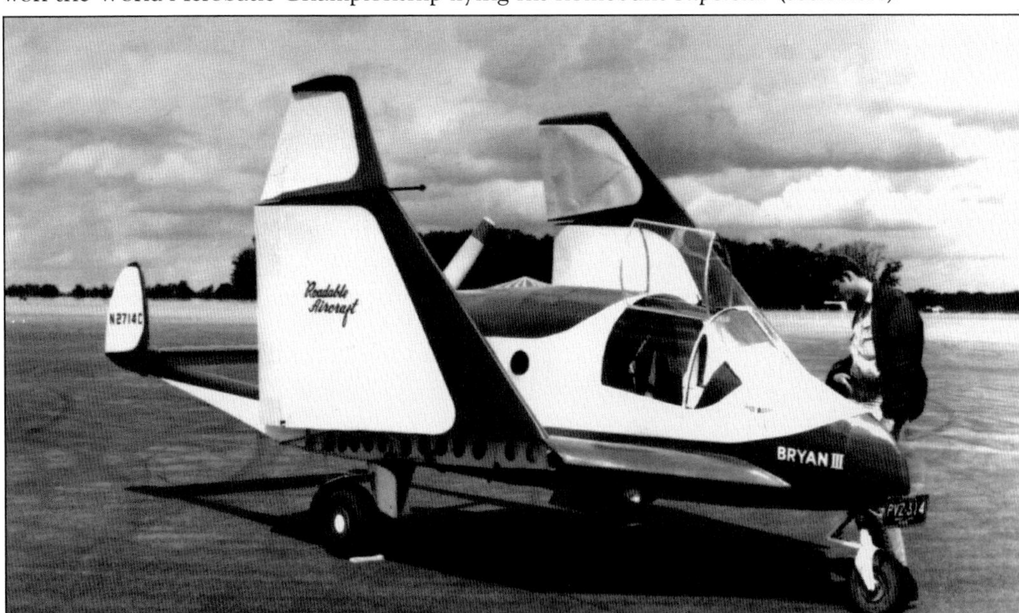

Dewey Bryan of Milford, Michigan, was obsessed with the idea of a roadable airplane. Using parts from an Ercoupe he built a prototype in 1953. A second Autoplane, the model II, was demonstrated at EAA's fly-in at Rockford, Illinois, and also at the Detroit Auto Show. Bryan's model III (seen here) was a two-seat version of model II. Dewey Bryan lost his life in 1974 when the folding-wing's locking pin failed on a test flight.

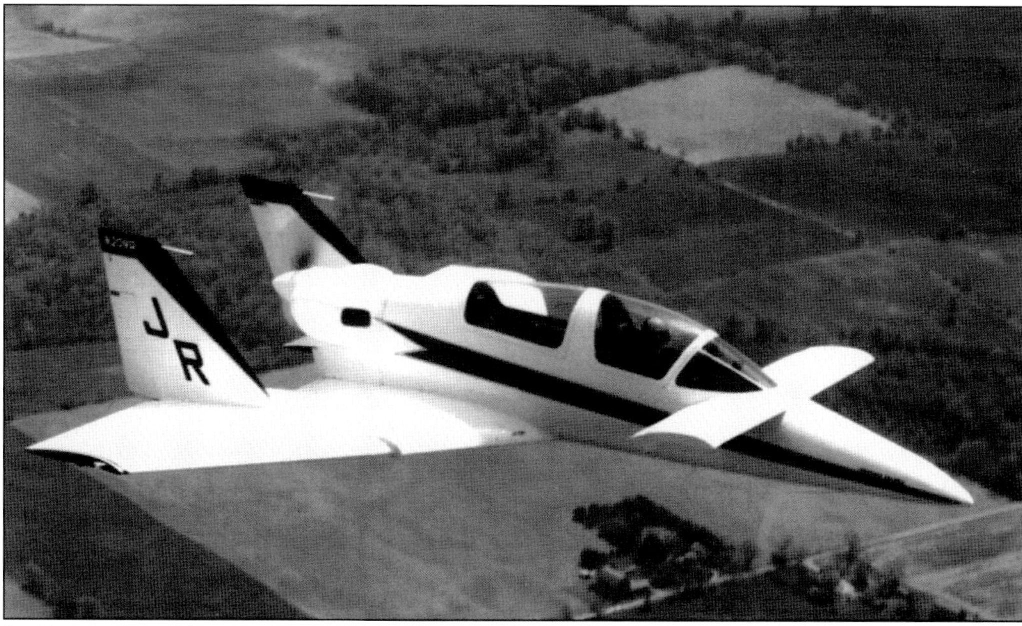

Joe Rayne flew Gloster Meteor jets in the British Royal Air Force. He came to Michigan in 1966 and built a single-place Midget Mustang, then bought plans for Burt Rutan's all-wood VariViggen. Rayne redrew those plans for all-metal construction, and the result was a much sleeker-looking airplane. The Viggenite was first flown in 1983. In 1989, Rayne had engine failure on takeoff and made an emergency landing. Wife Jeannie was unhurt, but Joe suffered major spinal injuries. (Photograph by the author.)

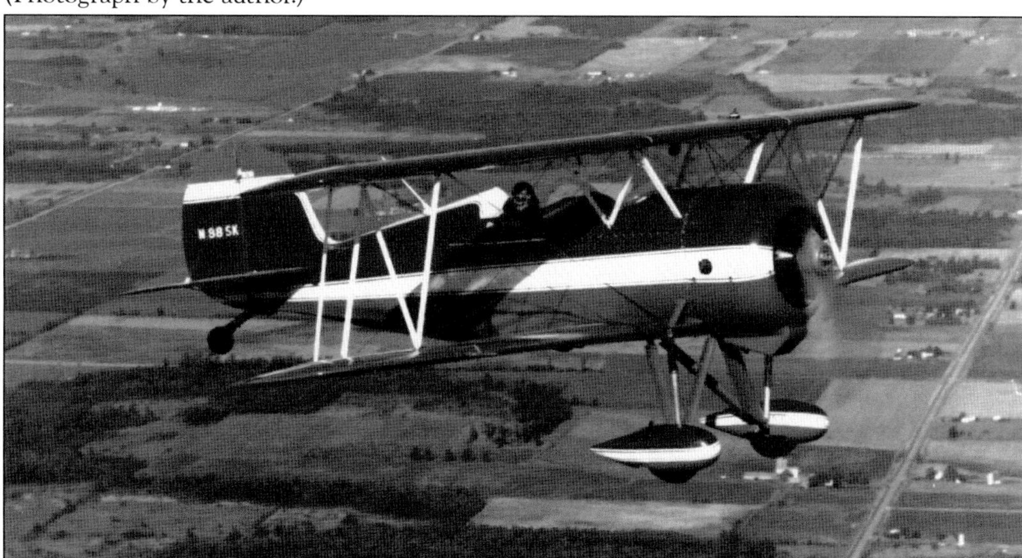

Dave McKenzie, born in Alabama, soloed at age 21. McKenzie's father, a former barnstormer, introduced him to aerobatics. After moving to Michigan, Dave became interested in the homebuilt aircraft movement and built this modified Great Lakes biplane, first flown in 1977. While performing aerobatics at an air show in July 1980 at Mason, Michigan, McKenzie executed a maneuver at too low an altitude and crashed. Although badly injured he eventually resumed flying. Now retired, McKenzie lives in Howell, Michigan. (Photograph by the author.)

AFTERWORD

Almost a century has passed since Michigan's first aircraft manufacturer, the Brooks Aeroplane Company, began building airplanes in 1911. Summing up the production numbers given with each aircraft described in this book, we arrive at a total of almost 30,000 aircraft built in Michigan since the pioneering days. The total includes a wide variety of flying machines ranging from lighter-than-air craft to multiengine transports and bombers and experimental airplanes of all types. Today only a small number of aircraft manufacturers remain active in Michigan. Enstrom Helicopter, in Menominee, has manufactured well over 1,000 helicopters. Classic Aircraft Corporation in Battle Creek has built 120 modern, updated versions of the classic 1934 Waco YMF-5 biplane. Staudacher Aircraft in Kawkawlin has built over 24 world-class Unlimited Aerobatic airplanes. And Mustang Aircraft in Troy has shipped hundreds of kits for single-place Midget Mustangs and two-place Mustang IIs to homebuilders around the world.

As we enter the second decade of the 21st century, the term "Michigan aircraft manufacturer" has fallen on the shoulders of those individuals who are presently building airplanes in their home workshops. With 6,200 EAA members in Michigan and 40 active local chapters, Michigan has a large number of potential aircraft builders. Each of those builders is following in the footsteps of early Michigan aircraft manufacturers like Eddie Stinson, Bill Stout, Fred Verville, and others, albeit on a smaller scale. At the present time, there are slightly over 7,000 aircraft registered in Michigan and approximately 1,400 are homebuilts. That number will surely rise as prices for factory-built aircraft continue to escalate.

It is difficult to predict what lies ahead in the amateur-built aircraft field. There has been talk of small, four-place turbine-powered homebuilts, and some have already appeared at EAA's AirVenture fly-ins. Looking back at the pioneering days of aviation it is amazing how much progress has been made in such a short time span. Aircraft design has progressed from primitive wood and fabric aeroplanes to supersonic aircraft, space travel and men on the moon. One can only speculate on what the future holds. Whatever it may be, much credit must be given to the early pioneers and aircraft manufacturers in Michigan and around the world who turned the mystery of flight into a reality.

BIBLIOGRAPHY

Allen, Richard S. *Revolution in the Sky*. Battleboro, Vermont: Stephen Greene Press, 1964.

Altman, Peter. *History of Aviation in the State of Michigan*. Lansing, Michigan: Michigan History Magazine, Vol. XXII, Spring 1938.

Ball, Robert S. *A Chronology of Michigan Aviation*. Lansing: The Inland Press, 1953.

Bell, Dana, editor. *Directory of Airplanes*. Washington, D.C.: The Smithsonian Air and Space Museum, 2002.

Biddle, Wayne. *Barons of the Sky*. New York: Simon and Schuster, 1991.

Bilstein, Roger E. *The American Aerospace Industry*. New York: Twaine Publishers, 1996.

Bluth, John A. *Stinson Aircraft Company*. Chicago: Arcadia Publishing, 2002.

Gunston, Bill. *World Encyclopedia of Aircraft Manufacturers*. Sparkford, Somerset, Great Britain: Haynes Publishing, 1993.

Ingells, Douglas J. *Tin Goose*. Fallbrook, California: Aero Publishers, 1968.

Juptner, Joseph P. *U.S. Civil Aircraft*. Nine vols. Fallbrook: Aero Publishers, 1961–1981.

Kidder, Warren B. *Willow Run*. Lansing: KFT Publishing, 1995.

Larkins, William T. *The Ford Tri-Motor*. West Chester, Pennsylvania: Schiffer Publishing Ltd., 1992.

Loving, Neal V. *Loving's Love*. Washington, D.C.: Smithsonian Institution Press, 1994.

Morrow, Walker C. *The Metalclad Airship ZMC-2*. Grosse Ile, Michigan. Privately printed, 1987.

O'Callaghan, Timothy J. *Henry Ford's Airport, 1909–1954*. Ann Arbor: Proctor Publications, 1993.

Pattillo, Donald M. *A History in the Making*. New York: McGraw Hill, 1998.

Peek, Chester L. *The Heath Story*. Norman, Oklahoma: Three Peaks Publishing, 2003.

Stout, William B. *So Away I Went!* New York: The Bobbs-Merrill Company, 1951.

Underwood, John W. *The Stinsons*. Glendale, California: Heritage Press, 1969 (Third printing 1982).

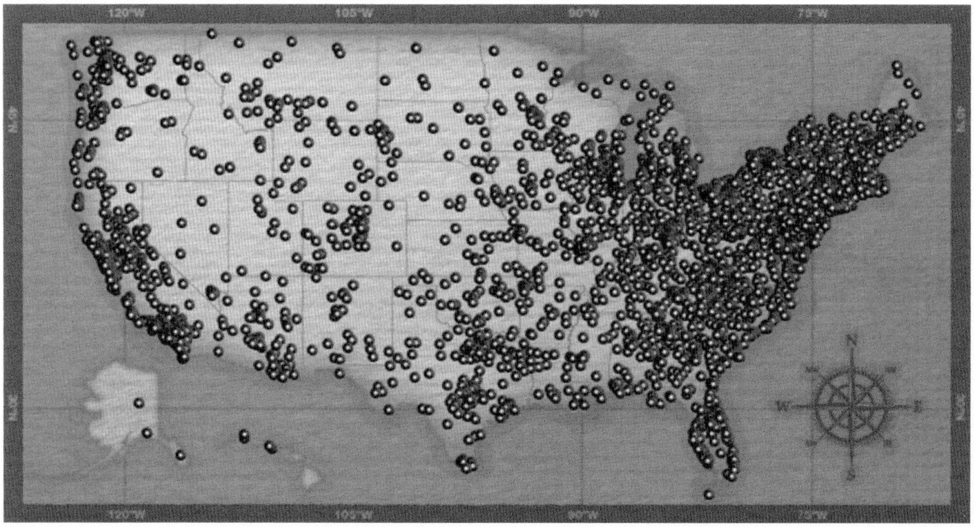